Marie C. Aguayo Beligat

# The Tree of Life and The Green Gold

***A crop that saves lives of humans and animals***

An amazing plant
that grows easily in arid and semi-arid areas
and can sustain life in adverse regions
producing
effective natural water biopurifiers,
functional foods with nutraceutical properties,
forage and feed animals,
and also
can be used for preventing, reverting and curing diseases,
fix desert and slopes, and improve soils

https://eidos-project.net
**NUTRI-SCIENCE FOR EVERYBODY**



ISBN 978-1-989332-06-1

The Tree of Life and The Green Gold.

Edition 2019.
Eidos-Project Books

# Table of contents

# PREFACE

"The Tree of Life and The Green Gold" is an amazing plant that can not only save human and animal lives, but can also help to maintain communities in inhospitable regions and solve the problem of hunger in many parts of the world.

It is a plant native to Mesoamerica that has peculiar characteristics that have been part of the Aztec empire and is now part of the food and nutritional culture of the Mexican population. After the discovery of America it was disseminated in the Mediterranean and other places of the planet. Since this plant can reproduce and propagate very easily, in many countries it is considered an invasive species and for lack of knowledge is not exploited or remains under exploited.

"The Tree of Life and The Green Gold" can easily grow without care or with very little care in arid and semi-arid areas and maintain life in communities established in adverse areas. This plant produces eco-friendly and efficient bioflocculants and biocoagulants that can be easily obtained and used, both at home and at an industrial level, for the natural biopurification of water —a basic and important element for the support and development of life everywhere— and produces functional foods with nutraceutical properties that they are able to prevent, to revert and cure various diseases including various types of cancer, produces fodder to feed animals such as cattle, pigs, goats and sheep as well as provides food for poultry, produces raw material for various kinds of food and non-food industries, is also a useful plant to be used in productive natural fences, as well as can improve and fix soils and slopes.

This book provides basic and general information about the multiple beneficial properties and the most important applications of "The Tree of Life and The Green Gold" that will allow the development of mini self-sustaining systems and obtaining economic benefits. It also provides ideas and knowledge to develop productive agricultural, medicinal and industrial farms with a positive impact on the local economy.

From "The Tree of Life and Green Gold" nothing is wasted, absolutely everything is taken advantage of.

# INTRODUCTION

The prickly pear tree is a plant that in addition to containing important phytochemical compounds, can grow not only in arid and semi-arid areas but can allow and sustain the development of life in adverse regions in the world. For its amazing characteristics it should be called "The Tree of Life and The Green Gold".

In an order of 1620, King Philip III of Spain states that: "*... one of the most valuable fruits grown in our Western Indies is the cactus pear, produce of equal value to gold and silver ...*"; in Ethiopia this plant is considered the '*Bridge of life*' because the stems and fruit store large quantities of water and provide both feed for cattle in times of drought and food for livestock herders, so contributing to the survival of both farmers and their animals; in Mexico for example, it may be possible to reduce malnutrition and to improve their quality of life[1].

In 1994, a Sicilian journalist described the virtues of "The Tree of Life and The Green Gold" and in a single phrase: "*A treasure that lies beneath the spines*"[1]. It is a vital plant that also has been called "*a crop that saves lives of humans and animals*" and in Latino America is considered "*the food of the poor*".

In Mexico pre-Hispanic times, "The Tree of Life and The Green Gold" was an important part of the Aztecs culture and had various uses such as food, drink, medicine, housing construction, fabrication of blankets and tools, and was important so large that it rose to a divine state[1]. Possibly "The Tree of Life and The Green Gold" was introduced to Europe in the middle of the sixteenth century by the Spaniard conquerors.

This amazing plant has a fast growth and produces phytomolecules that efficiently purify water —a fundamental element for the development of life anywhere in the world—, produces functional foods with nutraceutical properties with important medicinal properties, including anti-cancer properties, can be used as forage throughout the year for animal raising, can be planted to fix desert soils and slopes, and improve soils, can also be used to make productive fences, in addition to having multiple applications in the food and non-food industry. Juices, jams, flour, various types of preserves, jellies, dyes, sweeteners, biofloccu-lants for the bioclarification or biopurification of water are some of the products that it is feasible to elaborate from "The Tree of Life and The Green Gold". It is commercially exploited as fruit, vegetable, forage, energy, medicinal plant, pharmacology and natural colouring.

In addition, "The Tree of Life and The Green Gold" is a valuable plant that allows to establish important efficient and productive agro-ecosystems with socioeconomic benefits for the community ensuring additionally, satisfactory incomes in adverse environments. On the other hand, this plant can also be managed by sharing other crops in an efficient, productive and sustainable agricultural ecosystem.

In a sustainable agro-ecosystem, for example, both the "Tree of Life and Green Gold" and the soil can benefit from manure from animals fed mainly with cladodes and/or from a shared crop such as *Trifolium pratense* or red clover, a legume that is a beneficial forage plant that can fix nitrogen by increasing soil fertility. Natural fertilizers, from both manure and red clover, will benefit the cultivation of prickly pear by increasing plantation production and ensuring a variety of products used in the human diet, animal feed and providing raw material for a wide range of different types of industries.

At the same time, "The Tree of Life and The Green Gold" can play an important role in improving soil properties, can prevent avalanches, prevent soil erosion and flooding which are often factors responsible for natural disasters, improve soil moisture levels and water retention capacities to the benefit of agro-ecosystem productivity.

The good thing about "The Tree of Life and The Green Gold" is that it is a perennial crop that besides growing easily in adverse regions, requires only minimal or no care and tolerates a variety of environmental conditions, withstands intensive pruning and is also capable of efficiently managing the water resource.

Entire communities can be favored in different ways from the multiple benefits that "The Tree of Life and The Green Gold" can provide.

Nothing from "The Tree of Life and The Green Gold" is wasted, everything is taken advantage of.

# References

[1] Sáenz, Carmen et al. 2013. Agro-industrial utilization of cactus pear. Food and Agriculture Organization of the United Nations, FAO.

# CHAPTER 1

# BIOLOGICAL AND ECONOMIC IMPORTANCE AND CULTIVATION OF "THE TREE OF LIFE AND THE GREEN GOLD"

# BIOLOGICAL AND ECONOMIC IMPORTANCE AND CULTIVATION OF "THE TREE OF LIFE AND THE GREEN GOLD"

"The Tree of Life and The Green Gold" is a cactus known as a prickly pear or nopal. It is a plant with many useful and beneficial properties: it is a vital plant for the development of life in inhospitable conditions and has been qualified as "*a crop that saves human and animal lives*". The use and cultivation of *Opuntia* spp. dates back to prehistoric times. This plant can be easily cultivated and is part of the natural environment of many ecosystems. Prickly pear is a perennial crop that requires little or no care, withstands intensive pruning and tolerates a wide variety of environmental conditions.

## INTRODUCTION

"The Tree of Life and The Green Gold" is a cactus pear that is more than useful plant: it is a vital plant that has been called "*a crop that saves lives of humans and animals*", especially in times of severe drought[11]. Also, in countries of Latino America "*is the food of the poor*".

Cactus pears is *Opuntia* genus plant of the cactaceous family which is endemic in America; it is still the most widespread genus of the family, an one of the largest with 181 species and 10 naturally occurring hybrids[2]. *Opuntia ficus-indica*, with flattened stems segments belongs to the family Cactaceae that is reported to contain about 130 genera and nearly 1500 species, which is originally native to the New World; the prickly cactus pear (*Opuntia ficus-indica*) is widely distributed in Mexico, much of Latin America, South Africa and the Mediterranean area[18].

Cactus *Opuntia ficus-indica* has been commercially exploited as fruit, vegetable, forage, energy, medicinal and dye yielding crop in arid and semi-arid areas of world[12]. *Opuntia* spp. been as a source of food for the peoples of North America that the origins of its use is lost in ancient mythology, and in addition to food, it is used to treat whooping cough, diabetes, prostate problems, rheumatism, and nosebleed[19].

In add to be a good source of an eco-friendly effective natural water biopurifier, *Opuntia ficus-indica* —"The Tree of Life and The Green Gold" — is the cactus species of greatest agronomic importance, not only due to its delicious fruits, but also to its stems, which are used as fodder for livestock or as a vegetable for human consumption; the use and cultivation of *Opuntias* spp. dates back to prehistoric times, long before the Spaniards arrived in the Americas[11]. Unfortunately, in several parts of the world it has invasive species status and in those countries the potential beneficial uses of this plant remain unexploited or underexploited.

## "THE TREE OF LIFE AND THE GREEN GOLD" NAMES AND DESCRIPTION

| Scientific classification | |
|---|---|
| Kindom | Plantae |
| Clade | Magnoliopsida |
| Subclade | Caryophyllidae |
| Order | Caryophyllales |
| Family | Cactaceae |
| Subfamily | Opuntioideae |
| Tribe | Opuntieae |
| Genius | Opuntia spp. |

**Scientific name:** *Opuntia ficus-indica* (L.) Miller
**Synonymy**: *Opuntia maxima* Mill., *Opuntia gymnocarpa* Weber, *Opuntia ficus-barbarica* A. Berge.

**Fig. 1.1:** *Opuntia* spp. plant.
Prickly pear tree.
"The Tree of Life and
The Green Gold"

Flower.

**Fig. 1.2:** Prickly pear fruit, cactus pears or tunas.

**Fig. 1.3:** Tender young stems.
Nopalitos.

Fig. 1.4: Cladode with flowers and prikly pears. Flowers normally develop on the upper edge of the cladodes, up to 10 per cladode.

## Common names

Prickly pear tree, prickly pear, barbary fig, cactus pear, Indian fig, Indian fig prickly pear, mission fig, smooth prickly pear, tuna, cactus leaf, nopales, nopalitos, higo de pala, nochtli.

Nochtli is the original name, in the Náhuatl language.

## Description

The genus *Opuntia* includes the prickly pear, bunny ears, and beaver tail cacti; it is also called cactus pear and Indian fig[2]. The name most commonly applied to the plant was ficus indica, that is fig of the Indies[17]. The pads are called "cladodes" or "nopales" when they are whole, and "nopalitos" when they are diced or tender young cladodes. They taste something like green beans. The fruits are called prickly pears, cactus pears, or "tunas."

*Opuntia* spp., commonly are called nopal in Mexico, prickly pear, or cactus leaf in USA[6] , higo de pala in Spain and tuna in several countries of Latino America. A common use of cactus stems is nopalitos —tender young cladodes— a traditional vegetable eaten fresh or cooked in various dishes. Nopalitos are generally obtained from *Opuntia ficus-indica*, *Opuntia robusta*, or *Nopalea* spp.[8].

The flowers, which are sessile, hermaphrodite and solitary, normally develop on the upper edge of the cladodes; colour is variable and includes red yellow and white[13]. The fruit is an ovoid berry, big, with yellow color going to purple when ripe; the cladode weighs an average of 1.8 kg having about 40 cm in length, with rounded and ovoid shapes; most of the species from the genus *Opuntia* spp. have thorns involving the cladodes; the presence or absence of thorns is a characteristic often used for the identification of plant ecotypes; fruits vary from 4.8 to 10 cm long, with a width of 4 to 8 cm and weighing between 100 and 200 g; they have a thick rind that in the early stages of development is green, but this color will be changing to white-green, yellow, orange, red, purple, yellow-purple, violet or to brownish, depending on culture conditions[2].

Scientifically the stems are known as cladodes and commonly referred to as 'paddles', 'fleshy leaves' or 'racquets'; they are succulent and articulated; photosynthesis takes place in the cladodes, which act as leaves for this function; the cladodes are protected by a thick cuticle, which is occasionally covered with wax or glochids that further reduce water loss; the stems have the capacity to store considerable amounts of water, because they possess abundant parenchyma; this is a whitish tissue in which water is stored, enabling the plants to survive long periods of drought; cladodes also have spines; they have few stomata per unit of surface area and feature the peculiar property of remaining closed during the day and open during the night[13].

## ECONOMIC AND BIOLOGICAL IMPORTANCE

*Opuntia* spp. is a species of cactus with enormous economic importance throughout the world and due to its peculiar biological and economic characteristics deserves to be called "The Tree of Life and The Green Gold". *Opuntia* spp. plants are mainly a source of food for man and animals in the world due to the fact that they are available all year round.

Cactus pear was part of the Aztec culture[11]. In the sixteenth century *Opuntia ficus-indica* was already an important crop in central America; the Spanish took this species to their homeland, given its morphological peculiarities, and because of its edible fruits, anti-scurvy properties and for being the host of the cochineal insect, from which the commercially important cochineal dye was obtained; afterwards, from both México and Spain, *Opuntia ficus-indica* was introduced to other parts of the world, particularly to the Mediterranean region[9].

*Opuntia ficus indica* is the most agronomically important species for the production of edible fruits and cladodes[2].

In recent years, nopal has been recognized as one of the most economically important resources due to better valued nutritional, functional and therapeutic properties as well as for the potential it can have in the cosmetics and bioenergy industries, without forgetting the ecological applications and current nopal-based products of great value for current and future generations[3].

Cactus *Opuntia ficus-indica* has been commercially exploited as fruit, vegetable, forage, energy, medicinal and dye yielding crop in arid and semi-arid areas of world[12].

In the semi-arid northeastern of Brazil, palm is used in the combat to human hunger and malnutrition, and is a powerful alternative food rich in vitamins A, B complex, C and minerals like calcium, magnesium, sodium, potassium, as well as seventeen amino acids. It is a more nutritious food than cabbage, beets and bananas, in addition to the fact of being a cheap product[2].

In addition of a good source of food and medicinal properties, *Opuntia* spp. is also an effective and efficient eco-friendly natural water purifier for making drinking available water in the world.

## DISTRIBUTION, CULTIVATION AND PRODUCTION

The *Opuntia* genus, is a plant with about 1500 species of cactus[2]. While the prickly pear cactus is native to the United States, Mexico, and South America, it grows well in many areas of the world, including Africa, Australia, and the Mediterranean area and in some areas of South Africa and Australia, it has become a notorious weed; *Opuntia* species have a cosmopolitan distribution, found mainly in arid and semi-arid regions[2].

The countries where cactus is commonly cultivated are Mexico, Brazil, Argentina, South Africa, Israel, USA, Italy and many other Latin American countries, and cactus was identified as a potential crop for arid and semi-arid regions of world; possibly it is native from Mexico[12].

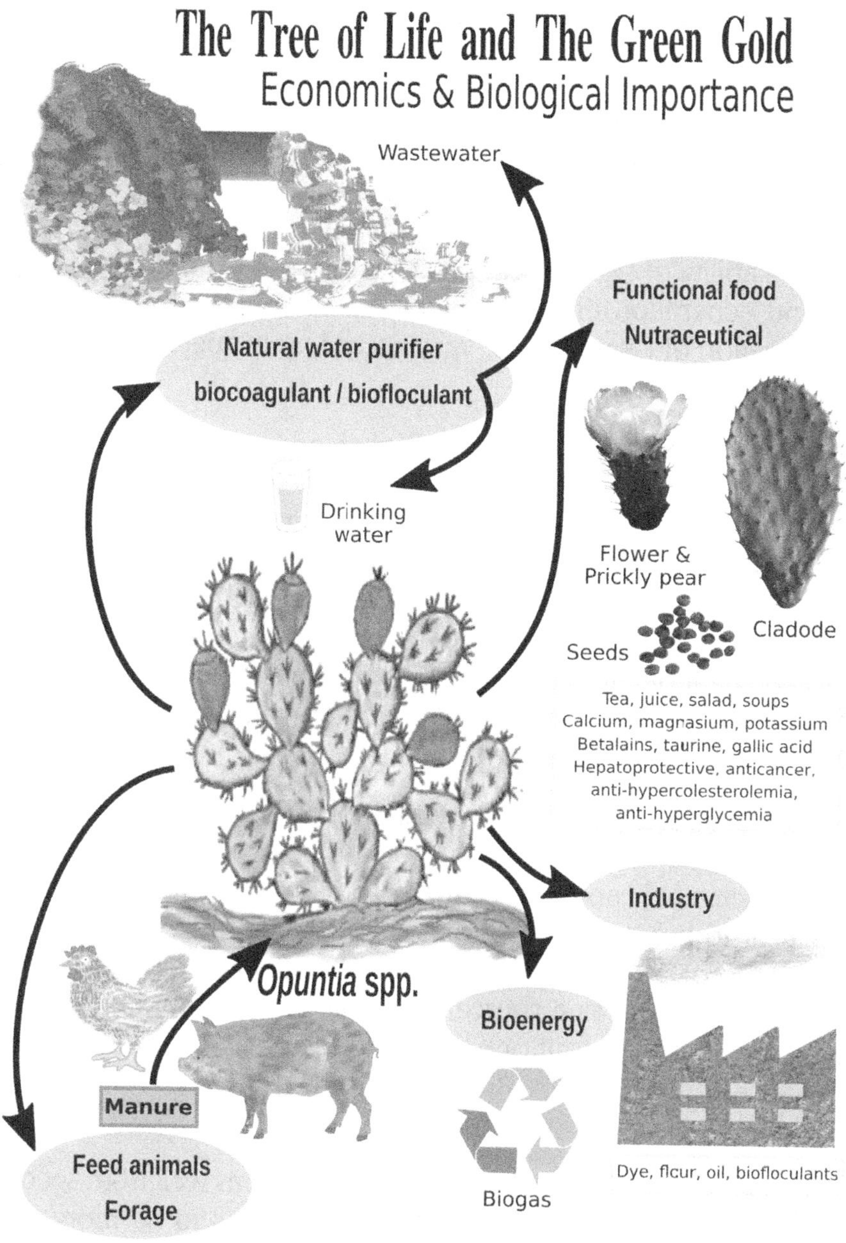


**Fig. 1.5:** Shows the multi economical and biological importance that "The Tree of Life and The Green Gold" has as a natural water biopurifier, source of food, agricultural, industrial and pharmacological applications.

Cactus pear is a perennial crop requiring minimal care. *Opuntia* tolerates a variety of growing conditions, but productivity in its natural habitat is limited by drought and poor soils; cladode shape has evolved to store the maximum amount of water with minimum loss; wild *Opuntia* spp. stands are usually found in poor soils with low contents of dry matter, in regions with a short growing season that does not allow the full expression of their growth potential; high inter-plant competition reduces the reproductive ability of *Opuntia* sp. plants, leading to extended juvenility and generation of new cladodes[1]. Of course, the chemical composition of *Opuntia* spp. vary according to species, ambient conditions and seasons.

The optimum conditions for its growth are available in summer rainfall regions having average rainfall between 300-600 mm; hot sunny days and cool dry winter where temperature do not fall below 5°C are most suitable for cactus production; cactus thrives best on sandy and sandy loam soils; however, it does well even on heavy soils with adequate drainage; gravelly or stony lands especially at the foot hill slopes are also suitable; further, it thrives well on slightly alkaline soils rich in calcium and potassium[12].

Cactus pear is easy to cultivate and has high productivity and is generally considered a rustic crop, able to survive in poor soils and dry areas[21]. Studies have shown that 1 hectare of *Opuntia* over 5 years old can produce up to 100 tonnes of fresh cladodes per year in areas with little rainfall (≤300 mm); in some semi-arid parts of Mexico, cladodes are traditionally collected from wild cactus plants as a source of forage; regular pruning boosts yield and improves fruit or nopalitos quality[11].

## Propagation

Propagation of *Opuntia* spp. can be started as well from seeds or from pads. Vegetative multiplication appears to be more efficient than sexual reproduction for plant recruitment[9].

For a faster production it is recommended propagation from pads because propagation from seeds is slow, and three to four

**Fig. 1.6 - 1.7:** Buds called aeroles at the side of the cladode developing new cladodes.

**Fig 1.8:** Three way for progragting *Opuntia* spp.
**1.** Up;
**2.** Angle of 30° and;
**3.** Lying
Settle it upright only about an inch deep in a mixture of equal parts of soil and sand or rough pumice; adventitious roots are developed from the areoles at the portion of the cladode making contact with the soil.

years may pass before flowers and fruits appear. Reproduction from pad is simpler and faster.

Planting material should be collected from robust, productive and healthy plants. The pads can be collected at the end of the growing season and subjected to slight dehydration to induce suberization of the joints. Collect pads of medium to large size, devoid of suspicious dark spots or discolorations; after collection, they are stored in a shaded dry place for 2 weeks[16].

The cladodes can initiate the rooting process soon after they come in contact with soil; the soil moisture is important –but not limiting– for rooting, because the root initials are supported by the water stored in the cladode[1].

That is the main reason that cactus pear is commonly propagated via cuttings; the stimulus to cell dedifferentiation and multiplication may occur within 2 days after the cuttings are placed in contact with the soil, and primordia emergence may occur within 14 days; from a growing cactus, cut a pad that is at least six months old, and allow it to form callous and after a month, some roots will have formed, and the pad will stand firmly by itself in the soil[8].

The buds, called areoles, can be found on both sides of the cladodes. These are capable of developing new cladodes, flowers or roots depending on environmental conditions[13]. During the first year, 2-4 pads per plant can be collected[16].

The plant can be maintained in the juvenile stage indefinitely by continuous pruning, which is the basic crop management tool for vegetable *Opuntia* spp. production; if not pruned, the cladodes will continue growing until autumn, giving rise to flowers at the beginning of spring; development of floral buds is mostly observed in mature cladodes that are at least six months old[1].

*Opuntia ficus-indica* and *Opuntia* spp. seeds have low germination capacity mainly due to their hard lignified integuments,

**Fig. 1.9:** Wild *Opuntia* spp. "The Tree of Life and The Green Gold".

**Fig. 1.10:** Wild *Opuntia* spp. "The Tree of Life and The Green Gold".

the most inward of which is the funiculus that envelops the embryo and obstructs radicle protrusion[2].

### Plantation

The matured cladode of identified accessions should be selected for planting; curing of freshly harvested cladodes is necessary under shade for 15 days before planting in the field; the cladodes should be treated with fungicide before planting; though the cactus can be grown at a close spacing of 2 m x 1 m, plant spacing of 3 m x 2 m was adopted for BAIF plantation; the cladode should be planted erect keeping 1/3rd portion below ground[20].

### Controling production

Production of *Opuntia* spp. can be controlled by varying the amount of nitrogen as fertilizers. If you prefer more flowers and fruits, give the plants a no-nitrogen fertilizer such as 0-10-10 (N-P-K) once a month, even through the winter. During this dormant period, the plants require a bright situation and enough water only to keep the pads from shriveling; but, if you want to grow the prickly pear for its pads, feed with a high-nitrogen fertilizer[7].

In rural areas, poultry and cow faeces can be used as a high-nitrogen fertilizer with a free cost.

### Harvesting pads

The best time for harvesting the pads is from mid-morning to mid-afternoon, when the acid content in the pads is at its lowest and remove the pads by carefully cutting them from their supporting pads[7].

### Harvesting fruits

Collection should be done with leather or rubber gloves and tongs; fruits, once harvested, **will not continue to ripen**, so they should be harvested ripe for the highest sugar content; use tongs to gently twist the fruit to separate it from the pad. It's best

**Fig. 1.11:** Flower and developing fruit.

**Fig. 1.12:** Opuntia Spp. roots.

**Fig. 1.13:** Cladode infected with cochineal insect. It is used for colorant production.

### TREATMENT FOR PADS MULTIPLICATION

To reduce rotting, the pads need to be treated with Bordeaux mixture prepared on the same day as treatment: mix 1 kg of copper sulphate in 5 litres of warm water until completely dissolved, then add 1 kg of lime, stirring until the mixture is homogeneous, and then dilute to 100 litres; this solution is enough to treat up to 2000 pads[16].

to put the fruit in a paper bag or some other container that will prevent the glochids from coming through[16].

### Harveting flowers

Collection of flowers should only be performed after the fecundation has taken place.

After drying the flowers in the shade, store the dried flowers in a cool, dry, and dark place.

## ANATOMY AND PHYSIOLOGY

Cactus stem is an ancestral vegetable that exhibits morphological, anatomical and physiological adaptations, which allow this vegetable to survive in a hostile environment[15].

Cacti have good water-use efficiency, thanks to the crassulacean acid metabolism (CAM) photosynthetic pathway and can withstand prolonged drought, high temperatures, and wind and water erosion; *Opuntia* spp. is utilized in programmes to prevent soil erosion and combat desertification; they are very adaptable, growing in severely degraded soil which is inadequate for other crops, and are ideal for responding to increases in atmospheric carbon dioxide ($CO_2$) levels; also, cactus can take advantage of the lightest rainfall, because its roots are close to the soil surface; the water is quickly collected by the roots and stored in thick, expandable stems[11]. In most cactaceae the photosynthetic tissue is in the stem and the stem tissues of cacti often contain large quantities of the complex carbohydrate mucilage, which is hydrophilic and affects water relations[8]. Also, the mucilage constitutes approximately 14% of the dry weight of the cladodes and can hold more than 30% of the total water of the reserve parenchyma[15].

Cacti possess a variety of physiological adaptations for growth and survival in desert environments; these plants have CAM, which is characterized by the occurrence of nocturnal transpiration and $CO_2$ assimilation, and the continuous day/night reassimilation of endogenous gases due to strict stomata regulation that enhances water use efficiency, while reducing the loss of endogenous gases and also by the continuous day/night reassimilation of endogenous gases that maintains a moderate level of metabolic activity during periods without atmospheric gas exchange[14]. The exchange of gases with the surrounding air occurs throughout the stomata, the only physical openings in the epidermis and the aperture of each stoma is controlled by two guard cells; stomata frequencies for cacti are low[8].

The growth buds become structures called areolas from where flowers, fruits and thorns are developed; the latter provide protection to avoid predation as well as damage from prolonged exposure to sunlight; thorns are modified leaves of the cactus and these are composed of cellulose (47.9 %), hemicelluloses (48.4%), lignin, ashes, fat and waxes being in very small amounts; additionally, a cuticle layer covers the surface of each thorn providing protection against desiccation; the hydrophobic layers are stacked and shifted circularly because of the growth of the spine from the cladode, leading to a thin section at the extremity and a larger one near the cladode[15].

## References

[1] Mondragón-Jacobo, Candelario et al. 2014. Opuntia as forage Cultivation of opuntia for fodder production: from re-vegetation to hydroponics. https://www.researchgate.net/publication/265260985.

[2] Guedes Paiva, Patrícia Maria et al. 2016. Opuntia sp. Cactus: Biological Characteristics, Cultivation and Applications. AIR, 7(3): 1-14.

[3] Ciriminna, Rosaria et al. The unfolding bioeconomy of nopal (*Opuntia* spp.)

[4] Ciriminna, Rosaria et al. 2019. Solar air drying for innovative *Opuntia ficus-indica* cladode dehydration. 4open, 2,1.

[5] Vijayaraghavan , G. et al. 2011. Application of plant based coagulants for waste water treatment. IJAERS, Vol. I, Issue I, October-December, 2011/88-92.

[6] Miller, Sarah M. et al. 2008. Toward Understanding the Efficacy and Mechanism of *Opuntia* spp. as a Natural Coagulant for Potential Application in Water Treatment. Environmental Science & Technology / Vol. 42, No. 12.

[7] Small Farm Center, University of California. 1989. Prickly Pear Cactus Production. http://sfp.ucdavis.edu/pubs/brochures/Pricklypear/

[8] Nobel, Park S. 2001. Cacti. Biology and uses. University of California Press.

[9] Reyes-Agüero, J. Antonio et al. 2005. Systematic notes and a detailed description of *Opuntia ficus-indica* (L.) Mill. (Cactaceae). Publicado como Ensayo en Agrociencia 39: 395-408.

[10] Miller, Sarah M. et al. 2008. Toward Understanding the Efficacy and Mechanism of Opuntia spp. as a Natural Coagulant for Potential Application in Water Treatment. Environ. Sci. Technol, 42, 4274–4279.

[11] Inglese P, et al. 2017. Crop ecology, cultivation and uses of cactus pear. FAO.

[12] Gajender, R.K. Yadav et al. 2013. Growth and fruit characteristics of edible cactus (*Opuntia ficus-indica*) under salt stress environment. Journal of Soil Salinity and Water Quality 5(2), 136-142.

[13] Sáenz, Carmen et al. 2013. Agro-industrial utilization of cactus pear. FAO.

[14] Szarek, Stan R. and Ting, Irwin P. 1974. Respiration and Gas Exchange in Stem Tissue of *Opuntia basilarist*. Plant Physiol. 54, 829-834.

[15] Ventura-Aguilar, R.I. et al. 2018. Cactus stem (Opuntia ficus-indica Mill): Anatomy, physiology and chemical composition with emphasis on its biofunctional properties. doi: 10.1002/jsfa.8493.

[16] Bishop, Carol and McMurray, Brynne. 2016. Eating Cactus: Prickly Pear for Food. University of Nevada. Cooperative extension.

[16] Mondragón-Jacobo, Candelario et al. 2001. Cultivation of opuntia for fodder production: from re-vegetation to hydroponics. FAO.

[17] Donkin, R. A. 1977. Spanish Red: An Ethnogeographical Study of Cochineal and the *Opuntia* Cactus. JSTOR.

[18] Kaur, Manpreet et al. 2012. Pharmacological actions of Opuntia ficus indica: A Review. Journal of Applied Pharmaceutical Science 02 (07): 15-1.

[19] Anderson, Edward F. 2001. The Cactus family. Timber Press. Portland, Oregon.

[20] Kauthale, Vitthal et al. 2017. Cactus an Emerging Fodder Crop of Arid and semi Arid India. BAIF Development Research Foundation.

[21] FAO. 2017. Crop ecology, cultivation and uses of cactus pear. Food and Agriculture Organization of the United Nations and the International Center for Agricultural Research in the Dry Areas.

# CHAPTER 2

# WATER FOR LIFE

## 2A

### "THE TREE OF LIFE AND THE GREEN GOLD"
AN EFFECTIVE NATURAL WATER BIOPURIFICATION

## 2B

### OVERVIEW OF WATER RESOURCES

# 2A

# "The Tree of Life and The Green Gold" is an effective natural water biopurification

# "THE TREE OF LIFE AND THE GREEN GOLD" IS AN EFFECTIVE NATURAL WATER BIOPURIFICATION

Today, various types of plant and cladodes extracts can be used as eco-friendly natural water biopurifiers or biocoagulants/bioflocculants. "The Tree of Life and The Green Gold" is a plant that can easily grow under difficult environmental conditions. It allows to easily obtain products for the efficient treatment of domestic and industrial wastewater and the obtaining of good quality drinking water. This plant can provide abundant natural water purification products that are cost effective, safe for human health and biodegradable. "The Tree of Life and The Green Gold" is a cactus with stems in flattened segments called cladodes or pads whose fruits are known to people as prickly pear or tuna. It is a surprising plant that has not only been used as a source of food by the peoples of North America, but in addition to its use as a natural biopurifier of water, is exploited commercially as a fruit crop, horticultural, forage, energy, medicinal and producer of natural colorants.

## NATURAL WATER BIOPURIFICATION

The natural biopurification of water is known as biocoagulation and bioflocculation, these processes involve macromolecular biocoagulants and bioflocculants biodegradable compounds secreted either by microorganisms or plants, usually these biocompounds show biodegradability, safety and absence of secondary pollution; bioflocculants are widely useful in the treatment of all types of water such as wastewater and those polluted waters resulting from the processing and manufacture of food and chemicals; biocoagulants and bioflocculants are known for their ability to clarify water turbidity[14].

Natural water biopurification or biocoagulants/bioflocculants have been used for domestic household for centuries in traditional water treatment in tropical rural areas; some reports describe natural biocoagulants from Nirmali seed —*Strychnos potatorum*—, maize, mesquite bean —*Prosopis* spp.—, *Cactus latifaria*, *Cassia angustifolia* seed and different leguminose species, but the material which has recently received the greatest degree of attention is the seed of *Moringa oleifera* indigenous to Sudan[5], and also *Opuntia* spp. as an effective and a friendly eco-bioflocculant.

Natural biopurifications of water with biocoagulants/bioflocculants are preferable to the **application of synthetic polymers and inorganic coagulants/floculants because the presence of residual monomers is undesirable considering their neurotoxicity and strong carcinogenic properties**; as it is well known, aluminium salts is the most widely used inorganic coagulant in water and wastewater treatment in the world. Aluminum sulfate (alum), a common coagulant used worldwide in the treatment of natural and wastewater waters, can achieve 90–99% microbial removal under optimal condition; however, alum produces large volumes of sludge and reacts with alkaline waters leading to a reduction in pH and shows low coagulation efficiency in water at low temperatures; in addition, alum has raised a number of concerns, including: 1. Ecotoxicological impacts when introduced into the environment as post-treatment sludge; 2. Impacts on human health as a result of the consumption of treated water; and 3. The cost of importing these chemicals for developing communities[10].

The use of natural water biopurification, biocoagulants and bioflocculants, it is not a new technology, for example there are Sanskrit writings from India that reported that the seeds of *Moringa oleifera* were used to clarify turbid surface water over 4000 years ago which indicated that they were the first reported plant-based biocoagulant used for water treatment[5]. With regard to special applications, actually, farmers in Chile and some other countries use cactus mucilage to clarify drinking water; as for other water-

# THE SOURCE OF THE NATURAL WATER BIOPURIFICATION

**Fig. 2A.1:** *Opuntia ficus-indica* flower.

**Fig. 2A.2:** Cladode with prickly pears.

**Fig. 2A.3:** flower and prickly pear.

**Fig. 2A.4:** *Opuntia ficus-indica* "The Tree of Life and The Green Gold"

soluble polymers, mucilage flocculates sediment particles and precipitates them out of solution[8].

## A GREAT NATURAL WATER BIOPURIFICATION

Natural biopurification of water with biocoagulants/bioflocculants from local plants are the most suitable products, mainly for obtaining homemade natural biopurifiers of water, being particularly useful in populations where people do not have public water purification systems.

Natural biopurification of water with biocoagulants/bioflocculants, such as *Opuntia* spp. and *Moringa oleifera*, may be part of an appropriate approach to household water treatment; this technology is based on local materials and labour, renewable resources and plant materials that fall within the local food category and is also relatively inexpensive to obtain and can contribute to advancing the goal of sustainable wastewater treatment technologies; *Opuntia* spp. is a plant with great natural biopurification power of water and is an ideal plant because it can grow in arid and water-scarce regions where *Moringa oleifera* is not found indigenously[7].

The prickly pear or "Tree of Life and The Green Gold" is a plant that tolerates a variety of soils, temperatures and humidity levels. Water purification or turbidity reduction by *Opuntia* spp. is as effective as the use of *Moringa oleifera*. *Opuntia* spp. is a good source of raw material for the natural biopurification of water, mainly *Opuntia ficus-indicates* that it is the most cultivated cactus species in the world.

The prickly pear is a plant, as mentioned in the previous chapter, that belongs to the genus *Opuntia* of the family of cacti that are endemic in America[2] and the prickly pear *Opuntia ficus-indica* is widely distributed in Mexico, much of Latin America, South Africa and the Mediterranean area and other parts of the world[4], that is to say that it is within the reach of many peoples dispersed throughout the world.

The juice or sap of the cladiodes or mucilage of the *Opuntia ficus-indica* cactus that it can be obtained very easily, it can be used as an effective and ecological natural biocoagulant/bioflocculant; this biocoagulant/bioflocculant in addition to being easy to obtain, is cheap and easy to apply for the treatment of both polluted natural water and industrial wastewater to obtain drinking water. It was discovered that cactus produces a gummy mucilage that can adhere to certain metals such as arsenic, nickel and chromium present in water samples; *Opuntia ficus-indica* mucilage is a promising phytocomponent in the field of emerging technologies for the removal of metals, and microbes[10].

On the other hand, in addition to its usefulness to obtain drinking water, it is a plant that has multiple benefits since the nopal *Opuntia ficus-indica* has been commercially exploited as a fruit crop, horticultural, forage, energy, medicinal and producer of natural dyes in arid and semi-arid areas of the world[3] and has also been a source of food for the peoples of North America and the origins of its use are lost in ancient mythology[1].

## HYDROCOLLOID PHYTOCOMPOUNDS

The genus *Opuntia* (v.n. "Nopal" Cactaceae) is characterized by the production of a hydrocolloid commonly known as mucilage, which forms molecular networks that are able to retain large amounts of water; the mucilage structure is proposed as two distinctive water-soluble fractions; one is a pectin with gelling properties with $Ca^{2+}$, and the other is a mucilage without gelling properties[9]. The cactus cladodes are mainly constituted by heteropolysaccharides with a molecular weight from $2.3.10^4$ to $3.10^6$ g/mol[5].

Mucilage is a gummy substance produced in cells found both in the chlorenchyma and parenchyma that helps cactus to retain water; several groups have reported that the mucilage of *Opuntia* spp. is a mixture of acidic and neutral polysaccharides consisting primarily of L-arabinose, D-galactose, L-rhamnose, D-xylose, and galacturonic acid[5,7].

The Fourier-transform infrared analysis indicates that dried *Opuntia ficus-indica* surface contains a variety of functional groups such as carboxyl, hydroxyl, sulphate, phosphate, aldehydic, ketonic, amine and alkyl groups. On these bases, it may be inferred that this biomaterial can give good results in terms of efficient reduction, biocoagulation/bioflocculation and biosorption of pollutants from wastewaters[8].

*Opuntia ficus-indica* cladodes can be used either as fresh plant parts or as dry powdered material to water biopurification treatments. Natural water purification or biocoagulants/bioflocculants are considered environmentally friendly due to their biodegradability, stability and low cost; it is forecasted that their use will promote more biodegradable sludge at the end of the process[8].

## HOMEMADE NATURAL WATER BIOPURIFICATIONS

Homemade water purification or biocoagulants/bioflocculants preparations **do not** require specialized techniques and special tools. Steps and methods for getting liquid and powder water biopurification or biocogulants/bioflocculants from cactus cladodes are easy and simples.

## EXPERIMENTAL PROCEDURES FOR USING WATER BIOPURIFICATION FOR CLEANING WASTEWATER

Efficiency and mechanism of biocoagulation-bioflocculation process depend on several factors, the most relevant being initial turbidity, pH, reagents (coagulant, adjuvant) dosage and type, system hydrodynamics in biocoagulation and bioflocculation stages, temperature, and alkalinity[11].

### THE FIRST STEPS PROCEDURES

- After to collect fresh cactus pads, the first step to be done, it is to remove thorns from cladodes.
- Cladodes must be cleaned using fresh water.

## Homemade natural water biopurification or Biofloculant in Mucilage[11]

- Take **1 kg** cactus cladodes *Opuntia* spp.
- Cut into small pieces 1x1 cm. using a kitchen knife to make easy the mucilage extraction in the boiling process.
- Add **500 ml** of water in a glass bottle plus the pieces of cactus.
- Heat at **85°C for 20 minutes** for inactivating enzymes. The temperature can be controlled in a Bain-Marie.
- Leave cool at room temperature.
- Filter the gum solution using a strainer.
- Adjust to pH 7.0.

## Homemade natural water biopurification or Biofloculant in Powder[5]

- Collect cactus cladodes.
- Wash repeatedly with water to remove thistles and dirt particles.
- Dry the cladodes under the sun for 3 hours.
- Cut into small pieces.
- Powder the small pieces using a domestic mixer.
- Dry at 60°C during 24 hours.
- Store the bio-agent in a glass bottle for further use agent without any pre-treatment

## Adjusting bioflocculants

After obtaining the biocoagulants/bioflocculants from the cactus to perform the biopurification of the water —liquid or powder— it is important to find the optimum dose of the biocoagulant/bioflocculant as well as the best pH value for the wastewater in order to obtain the maximum efficiency of the biocoagulant/bioflocculant of the cactus.

1. Vary the pH with values above 7.0 until to find the best bioflocculation. Put samples in glass bottles and try different pH for each bottle and determine the best pH value.
2. Vary the dose of biocoagulant between 0.5 and 5% in order to ascertain the influence on the biocoagulation-bioflocculation process and to determine the optimal dose.
3. Evaluate the best time of the flocculation process.

## Biofloculant experimental dosage

The optimum dose of a bioflocculant is defined as the value which there is no significant increase in removal efficiency with further addition of bioflocculant.

Optimum dose of *Opuntia* spp. ranges from **5 to 15 mg/L in low turbidity water, 15–35 mg/L in medium turbidity water, and 35–55 mg/L in high turbidity water**.

*Opuntia* spp. **operated with greater than 98% turbidity removal from pH 8–10**. Scientific findings have reported that *Opuntia* spp. is most effective at pH 10 and is least effective at pH 6[7]. pH can be adjusted using a solution of sodium hydroxide (NaOH) or bicarbonate.

As the pH increased, the number of negatively charged sites increases, with consequent enhanced biosorption of positively charged metal[8] cations through electrostatic attraction forces.

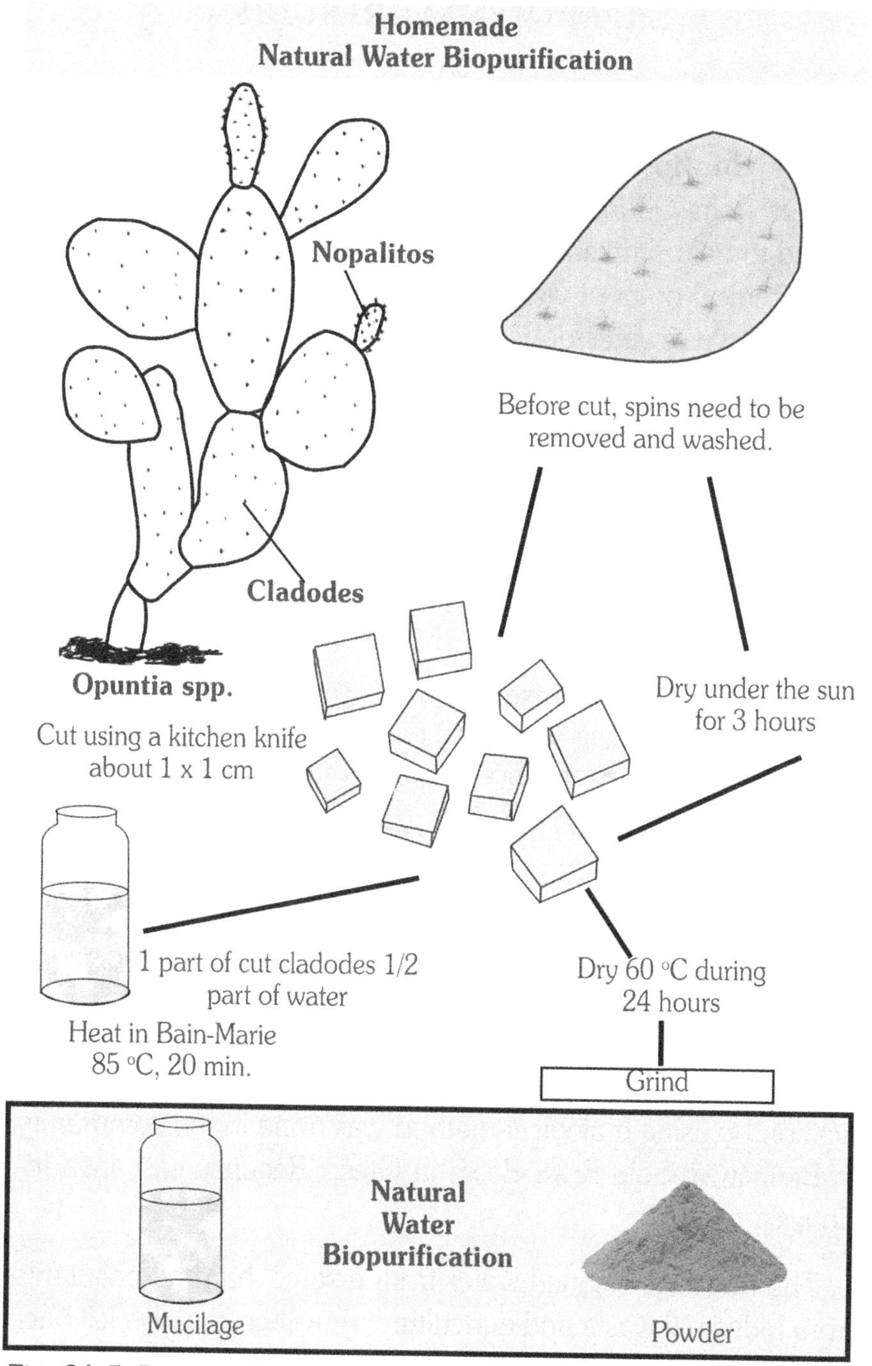


**Fig. 2A.5:** Processes for obtaining water biopurifiers from the cladodes of *Opuntia* spp.

# EXPERIMENTAL RESULTS

## Experiment 1

Using **bioflocculants powder**[5]: mucilage bioflocculants powder from *Opuntia ficus-indica* for pollution removal, was proved to be a simple and efficient treatment from an economic and technical point of view. From the bioflocculation process of industrial wastewater done in the field, the result for discoloration was equal to 99.89%, chemical oxygen demand removal of about 89.25% and turbidity abatement above 92.18% with an overall desirability value equal to 99.4%. In this experiment, the comparative study with commercial flocculants (EPENWATE EXP 31/1, Polyacrylamide A100PWG) showed that the *Opuntia ficus-indica* mucilage had a better performance with the following values.

- **Bioflocculants dose: 40mg/L**. The most effective dose of the bioflocculants was found to be 40mg/L at which the higher flocculation performance was observed.
- pH of the effluent about 7.2.
- Bioflocculation mixing speed of about 30rpm.
- Bioflocculation mixing time of about 11min.

## Experiment 2[6]

**Bioflocculant mucilage**: The experiment of water biopurification by using biological method was done in the community in Bengawan Solo River Basin in Gresik Regency east Java Indonesia.

The mucilage cleaned water from arsenic, harmful pollutants from industrial waste and agriculture; mucilage can also kill bacteria that was one of the harmful pollutants in drinking water, mucilage can produce clean water 98 percent free of bacteria.

**Finding the optimum dosage for purifying the water**

- Fill several glass bottles with 1 liter of water sample.
- Treat each sample with different dosages (mg or ml per L) of your *Opuntia* spp. homemade product. Try 5, 10, 15, 20, etc. until to remove the turbidity and find the best flocculation.
- Check the turbidity of the purified water using a turbidity tube.

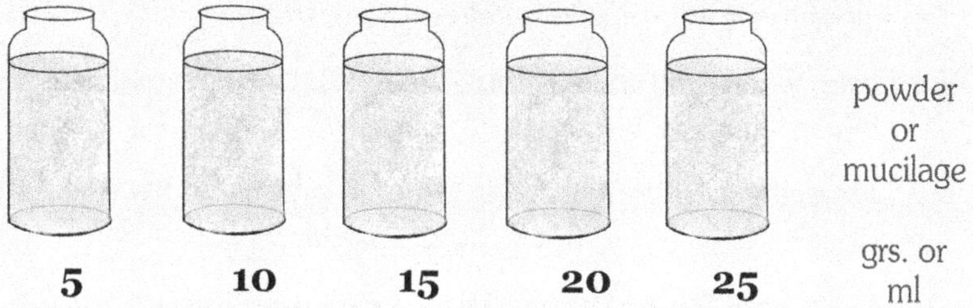


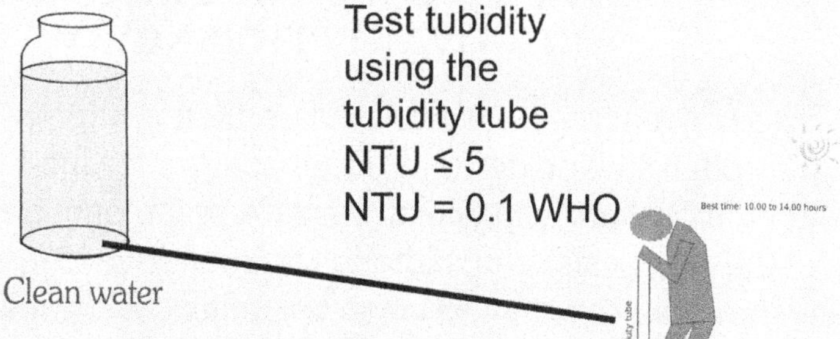


- Repeat the process using the previous best results, but modifying the pH of the samples.

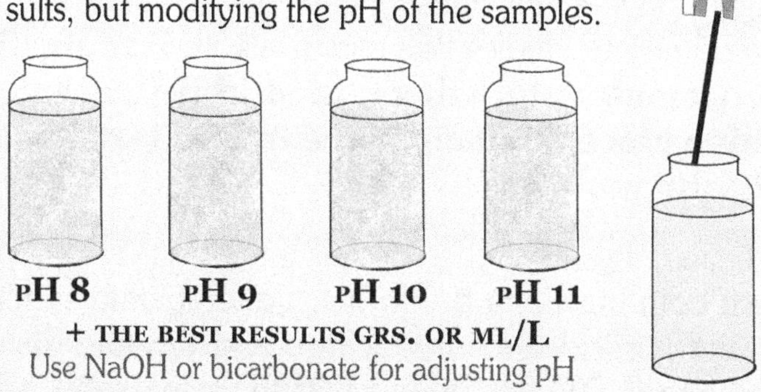


+ THE BEST RESULTS GRS. OR ML/L

Use NaOH or bicarbonate for adjusting pH

**Inadequate dosage of bioflocculant will lead in low flocculating activity; excess input might induce restabilization of particles.**

**Fig. 2A.5:** Processes for obtaining water biopurifiers from the cladodes of *Opuntia* spp.

### Experiment 3

**Removing contaminants from a wastewater sample**: Water treatment can be done individually or collectively. Steps and procedures[6]:

- Add cactus gum solution at a concentration of 5 percent of 25 ml and stir up.
- Allow to stand approximately one minute.
- Dirty water within 5 minutes can already be processed into clean water.
- Dump the clean and purifier water in a bottle, if necessary use a coffee filter to avoid flocks go in a clean water.

## WATER CONTAMINANTS REMOVAL EFFICIENCY

The conventional coagulation/flocculation process and techniques offer advantages and disadvantages for the treatment of both industrial and municipal wastewater; among the advantages of the chemicals used, although less sensitive to toxic charges, they are more sensitive to larger quantities of organic products, are easy to operate and save energy, among other properties; among the disadvantages of using them is that they **have been shown** in numerous studies that chemicals, such as **aluminum salts, acrylamides and others, used in the coagulation/ flocculation process remain in the treated water and can induce health problems**[12].

In particular, the researchers tested the efficiency of *Opuntia ficus-indica* both fresh and dry in batch and column systems to eliminate the pesticides aldrin, dieldrin and dichloro-diphenyl-trichloroethane (DDT) and found that the remarkable adsorption power of the pesticides possessed by *Opuntia ficus-indica* dry and fresh apparently depends on the size of the bioparticles and the adsorbent biomaterials properties; substantially smaller bioparticles allow excellent pesticide removal; for example, DDT is eliminated in the range up to 99.2% with the presence of bioparticles smaller than 0.25 mm in diameter, while bioparticles of 1.0-2.0 mm can eliminate up to 77.1% for this pesticide[8].

According to the conclusions of scientific investigations carried out, cactus mucilage is an effective biocoagulant/bioflocculant according to the results obtained in the tests with Dice-Cut-Pressed and the mucilage extracted from the cactus, both tests showed an almost similar adsorption capacity for heavy metals, total coliforms and fecal coliforms and the percentage of reduction of contaminants of both treatments was extremely effective and no significant differences were observed in the efficacy between the treatments; copper (Cu) and cadmium (Cd) showed a biosorption of 100% on the second day; lead (Pb), nickel (Ni), zinc (Zn) and iron (Fe) showed a higher adsorption for mucilage and fecal coliforms that were completely adsorbed also on the second day; the treatment protocol of cactus cubes was more efficient, reproducible and easy to use; according to the results of these tests, the mucilage of *Opuntia ficus-indica* showed to be a promising bioelement in the field of emerging technologies for the elimination of both metals and microbes[10].

Several studies indicated the importance of using cactus as a biocoagulant/bioflocculant agent for the removal of turbidity, chemical oxygen demand and heavy metals; in general, all studies focused on the preparation and optimization of the biopolymer based on added cactus, which can considerably affect the efficiency and cost of the process; biopolymer dosing is one of the most important parameters that must be optimized, since **insufficient dosing or overdosing would result in poor treatment performance**[15].

Interestingly, cactus is an abundant natural product, cost effective, safe for human health and it is biodegradable[12], that can be easily cultivated in a large range of area in the world.

## PADS DEHYDRATION

Powder from dehydrated *Opuntia*'s pads or cladodes is a functional ingredient widely employed in different commercialized products such as foodstuffs, nutraceutical dietary supplements, cosmetics products and powder bioflocculants for natural water biopurification.

Traditionally, nopal flour has been obtained by direct sun-drying of cactus cladodes sliced and left in the open environment for several days, followed by grinding the dehydrated cladodes and sieving; cladodes are difficult to dry because of the high mucilage content[13]. The pads industrial dehydration process needs to be done under controlled conditions in appropriate indoor processing facilities. Pads can be dehydrated by using natural gas as a burning fossil fuel or by using solar air heating.

Today solar crop drying systems are widely used across the world's sunniest countries to dehydrate crops improving crop product quality meeting the increasing demand of healthy natural foods at affordable cost; the advantage of the sun-dried cladodes is that they retain their original green color and throughout the whole matrix of the dried stems; using solar air heating and ventilation to dehydrate the fresh cladodes of *Opuntia ficus-indica* produces green dehydrated cladodes and green nopal powder and using solar air heating and ventilation to dehydrate the fresh cladodes of *Opuntia ficus-indica* produces green dehydrated cladodes and green nopal powder; and using a 36 $m^2$ solar array it is possible to dry 5 tonnes of fresh cladodes cut from the plants in a few hours[13].

## References

[1] Anderson, Edward F. 2001. The Cactus family. Timber Press. Portland, Oregon.

[2] López-Cervantes, Jaime et al. 2011. Functional properties and proximate composition of cactus pear cladodes flours. Ciênc. Tecnol. Aliment., Campinas, 31(3): 654-659.

[3] Gajender, R.K et al. 2013. Growth and fruit characteristics of edible cactus (*Opuntia ficus-indica*) under salt stress environment. Journal of Soil Salinity and Water Quality 5(2), 136-142.

[4] Kaur , Manpreet et al. 2012. Pharmacological actions of *Opuntia ficus indica:* A Review. Journal of Applied Pharmaceutical Science 02 (07): 15-1.

[5] Bouatay, F. and Mhenni, F. 2014. Use of the Cactus Cladodes Mucilage (Opuntia Ficus Indica) As an Eco-Friendly Flocculants: Process Development and Optimization using Stastical Analysis. Int. J. Environ. Res., 8(4):1295-1308.

[6] Reyshan. 2017. Installation of Water Purification and Using Cactus (*Opuntia cochenillifera*) As coagulant To Overcome The Clean Water Problem Around The River Community. Installation of Water Purification and Using. https://thebestscienceproject.com/making-installation-water-purification-miniature-using-cactus-opuntia-cochenillifera-coagulant-overcome-clean-water-problem-around-river-community/.

[7] Miller, Sarah M. et al. 2008. Toward Understanding the Efficacy and Mechanism of Opuntia spp. as a Natural Coagulant for Potential Application in Water Treatment. Environ. Sci. Technol, 42, 4274–4279.

[8] Barbera , M. and Gurnari, G. 2018. Wastewater Treatment and Reuse in the Food Industry, Chemistry of Foods. Chapter 4: Quality Standards for Recycled Water: *Opuntia ficus-indica* as Sorbent Material.

[9] Sepúlveda , E. et al. 2004. Extraction and characterization of mucilage in Opuntia spp. Environments Journal of Arid Environments 68; 534–545.

[10] Hassan, Sidra et al. 2017. Biosorption of heavy metals from sewage effluent using prickly pear cactus. J. Bio. Env. Sci. Vol. 10, No. 2, p. 16-24.

[11] Mounir , Belbahloul et al. 2014. Low Technology Water Treatment: Investigation of the Performance of Cactus Extracts as a Natural Flocculant for Flocculation of Local Clay Suspensions. International Journal of Engineering Research & Technology; Vol. 3 Issue 3.

[12] Rebah , F. Ben and Siddeeg, S.M. 2017. Cactus an eco-friendly material for wastewater treatment: A review. JMES, Volume 8, Issue 5, Page 1770-1782.

[13] Ciriminna, Rosaria et al. 2019. Solar air drying for innovative *Opuntia ficus-indica* cladode dehydration. 4open 2019, 2, 1.

[14] Mounir, Belbahloul et al. Comparison of the Efficacy of Two Bioflocculants in Water Treatment. International Journal of Scientific Engineering and TechnologyVolume No.3 Issue No.6, pp : 734-737.

[15] Rebah, F. Ben and Siddeeg, S.M. 2017. Cactus an eco-friendly material for wastewater treatment: A review. JMES, 8 (5), pp. 1770-1782.

# 2B

# Overview of Water Resources

# Overview of Water Resources

Water is a natural element absolutely essential for life; without water life is not possible. Water is a vital substance and of great importance in all natural and anthropogenic activities. More than a third of all deaths in developing countries are water-related, especially contaminated water, which is one of the leading causes of death and disease in the developing world. Wastewater pollution is one of the world's most important environmental problems and needs to find effective solutions for wastewater treatment. Water plays an essential role in our health, economy, food production and environment.

## INTRODUCTION

Water is as essential to us for living and to the Earth; it is the natural element that makes our lives possible. A person can survival three weeks without food but can't survival more three days without water.

Water is a vital substance and of great importance in all natural and anthropogenic activities and it is well know that water plays an essential role in our health, economy, food production, and environment; it regenerates shape and oceans and seas, rivers, lakes and forests, becoming part of the identity of environments and landscapes and of paramount importance for the development of ecosystems and human life[3].

It is a fact that the climatic changes affect the spatial and temporal distribution of the rains, being able to cause water deficit to the cultures; it should be noted that the increase in population and the consequent demand for larger quantities of food tend to increase the consumption of water resources[1].

Unfortunately 20% of the world's population or more than 1 billion people lack access to safe drinking water; 50% of the world's population or more than 3 billion people lack access to sanitation and about 80% of all illnesses; more than one third of all deaths in developing countries are related to water; it is estimated that worldwide, around 7 million die yearly from diseases linked to water; half of the world's rivers and lakes are seriously polluted and polluted water is the major cause of death and disease in the developing world[5].

In the world there are a lot of communities that do not have access to public water supplies, for example, populations that live in small communities and small villages usually do not have access to public water supplies and they obtains their water supply from unprotected sources such as open dug wells, streams and ponds which are polluted, so the treatment of water in these areas has a unique problem[3].

In recent years, increased industrial and agricultural activities and the correlated population growth led to over exploitation of natural resources; the production of large volumes of sewage has inevitably resulted in the increased generation of various types of pollutants; one of the consequences of this rapid growth represented by environmental disorders; the demand for clean and safe water has increased tremendously in different sectors; for these reasons, the hazardous pollution of wastewater is one of the most important environmental problems worldwide[4].

The hazardous pollution of wastewater is one of the most important environmental problems worldwide and need to find effective solutions for treating wastewater because water plays an essential role in our health, economy, food production, and environment.

Also, there is several sources of natural contamination of the fresh water, such as contamination of groundwater with arsenic from natural leaching of arsenic bearing minerals is an environmental phenomenon, which affects millions of people in different parts of world; according to the World Health Organization (WHO) recommendation is that the maximum concentration of arsenic in drinking water should not exceed 10 $\mu g/L$[10].

## WATER RESOURCES IN THE WORLD

While 70 percent of the earth's surface is water, only three percent of it, is fresh water, and almost all of that three percent is inaccessible for human use; about three-quarters of all fresh water on earth is locked away in the form of ice caps and glaciers located in polar areas far from most human habitation. In all, only about 0.01 percent of the world's total water supply is considered available for human use on a regular basis[8].

**The amount of water on Earth do not change. The amount of water cannot be increased or decreased. The amount of water on Earth always is the same.** Water on Earth moves through the hydrologic cycle. Instead, **fresh water is a renewable resource that his availability can vary place to place and also is distributed extremely unevenly around the globe**, but **only a small percentage of water on Earth is suitable for humans to drink**. With the increase of populations and industries there is increase of fresh water demand.

The water cycle is the movement of water through the Earth, but most of the fresh water on Earth is in storage; fresh water can

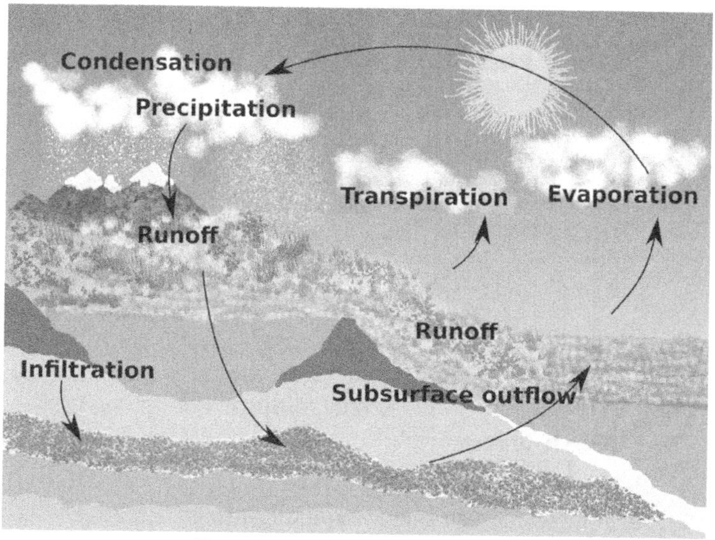


**Fig. 2B.1:** Hydrologic cycle.

**Fig. 2B.2:** Seventy percent of the earth's surface is water, but only three percent of it is fresh water and only about 0.01 percent of the world's total water supply is considered available for human use on a regular basis.

**THERE ARE ABOUT 1.1 BILLION PEOPLE IN THE WORLD WITHOUT ACCESS TO CLEAN WATER[11].**

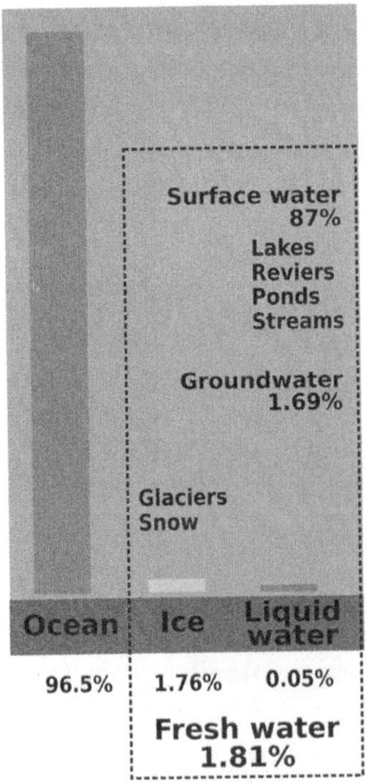

**Fig. 2B.3:** Access to clean water varies with geography, economics, politics, and culture; however, the worldwide community agrees that all of Earth's citizens deserve access to the planet's most essential resource[11].

**Fig. 2B.4:** Total water distribution on the Earth.
Source: https://en.wikipedia.org/wiki/Water_distribution_on_Earth

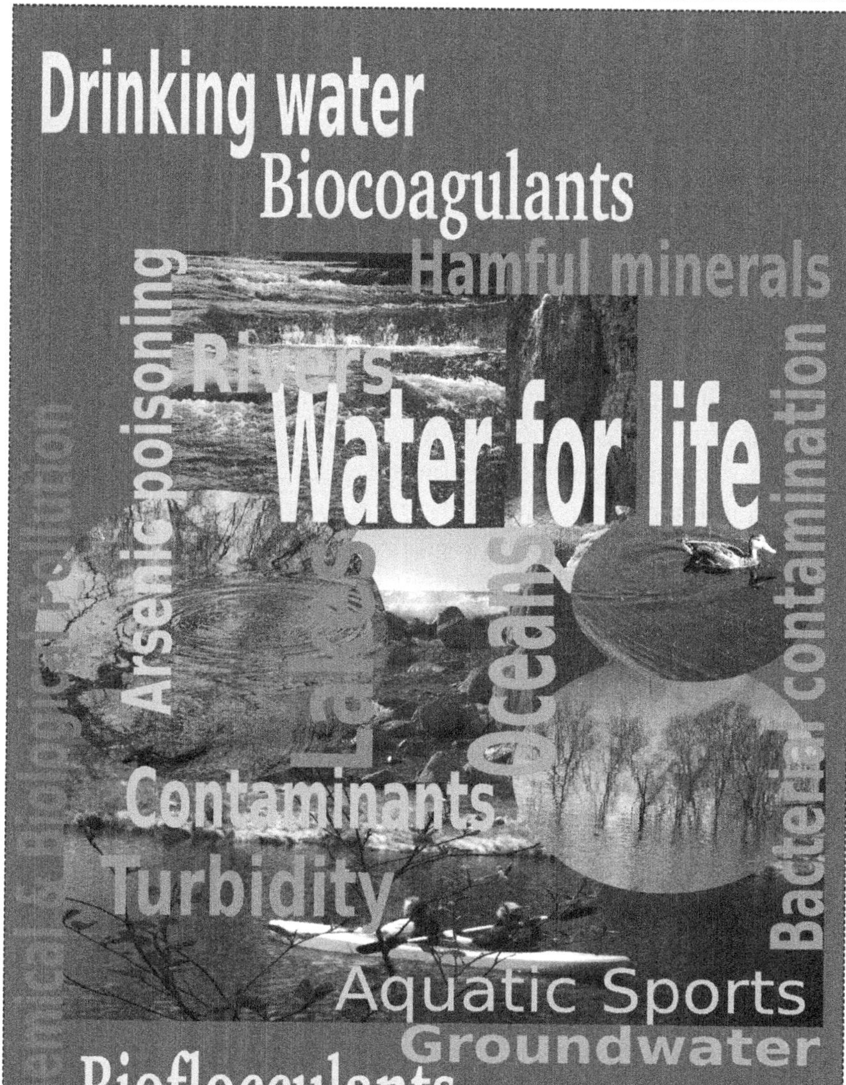


**Fig. 2B.5:** Water source can be polluted by various ways like chemicals, agricultural runoff and human and animal feces. Several plant species that can be used as a friendly and effective biocoagulant and disinfectant for getting safe and clean water.

be stored for days or weeks in a lake or for thousands of years underground or in an ice cap; Antarctica contains nearly 70 per cent of the world's fresh water as ice in an ice sheet that covers large sections of bedrock or solid ground in Antarctica[9].

## WATER TREATMENTS AND PURIFICATION

Drinking water treatments typically includes coagulation, sedimentation, filtration, and disinfection processes. Coagulation is a critical step in water treatment processes, not only because it removes particles, but because it is also removes the microorganisms that are often attached to the particles. That is, by removing turbidity, coagulants also have the potential to remove pathogens and to significantly improve water quality and, subsequently, human health[2]. After the purification, the water should be free of harmful microorganisms, and organic, and inorganic pollutants.

Flocculation is one of the important separation processes that are extensively employed in portable water, domestic or industrial wastewater treatment. Various types of flocculants have been developed and marketed for the removal of environmental concerned parameters such as suspended and dissolved solids, turbidity, chemical oxygen demand, colour and dye through sedimentation[12].

## WATER PURIFICATION

One of the inorganic coagulant for treating wastewater are aluminium salts that are 'cheap' and are the most widely used coagulants in water and wastewater treatment all over the world, but regarding the application of synthetic polymers, the presence of residual monomers is undesirable because of their neurotoxicity and strong carcinogenic properties; in that way there is strong **evidence that aluminium-based coagulants is linked to the development of Alzheimer's disease in human beings**[7].

For that reason, natural water biopurifiers plant-based called biocoagulants/bioflocculants emerge as an attractive alternative to polymeric coagulants/flocculants and their applications in wastewater treatment has become increasingly essential, because they are biodegradable, non-toxic and are also widely available from renewable resources, environmentally friendly and do not have a negative impact on the environment[12].

Some plant based biocoagulants/bioflocculants for water purification could be a good solution for populations in different parts of the world, especially biocoagulants/bioflocculants that are obtained from plants that can grow in deserts, in all types of soils and under several climatic conditions; in addition such plants can often also be used to control soil erosion without discounting that some of these plants can also be used at the same time as a good food source.

As mentioned in the previous chapter on water biopurification, there are communities around the world, especially rural communities, that have used and are using natural water biopurifiers, mostly from plants, to obtain safe, efficient and effective drinking water.

In addition to using plants that can be cultivated locally, they must allow to obtain and apply the biocoagulant/bioflocculant using homemade methods such as *Opuntia ficus-indica*. It is also important to be able to measure the level of contamination and clarification of the water, i.e. the turbidity of the water, using accessible and reliable measurement methods, such as the use of the Secchi disc.

## WATER TURBIDITY MEASUREMENT

Turbidity in water is a measurement of how cloudy or murky it is that is caused by particles suspended or dissolved in water that scatter light making the water appear cloudy or murky. Particulate matter can include sediment, especially clay and silt, fine organic and inorganic matter, soluble colored organic compounds, algae, and other microscopic organisms[13].

Turbidity is an important water quality parameter in drinking water provision and treatment and it is a key indicator used in assessing the suitability of water for human consumption[14]. The unit of measurement is called a Nephelometric Turbidity Unit (NTU); the greater the scattering of light, the higher the turbidity[13].

There are several methods that are available for testing turbidity in water. Turbidimeter (Nephelometer) and Secchi Disk[14] that are the instruments more used for measuring turbidity.

- **Turbidimeter (Nephelometer)**[14], is extremely accurate and it can measure very low turbidity. It has a beam of light that pass through water sample. It is expensive and can easily to be damaged. It requires a power source and calibration. Actually it is possible to buy a portable turbidimeter on Internet at a very reasonable price.
- **Secchi disk**[14], is a low cost and portable instrument and it is easy to learn and to build. It is less accurate and it can't be used in shallow water or swift currents and it is not applicable to small water sample size.

Drinking water should have a turbidity of ≤ 5 NTU[14]. The World Health Organization recommends that ideally, a median turbidity value of less than 0.1 NTU should be maintained for effective disinfection (WHO, 2004).

## SECCHI DISK

The Secchi disk, as created in 1865 by Angelo Secchi, is a plain white, circular disk 30 cm (12 in) in diameter used to measure water transparency or turbidity in bodies of water. The disc is mounted on a pole or line, and lowered slowly down in the water. The depth at which the disk is no longer visible is taken as a measure of the transparency of the water. This measure is known as the Secchi depth and is related to water turbidity. Since its invention, the disk has also been used in a modified, smaller 20 cm (8 in) diameter, black and white design to measure freshwater transparency; the original Secchi disk was a plain white disk and it was used in the Mediterranean Sea[15]. Actually we

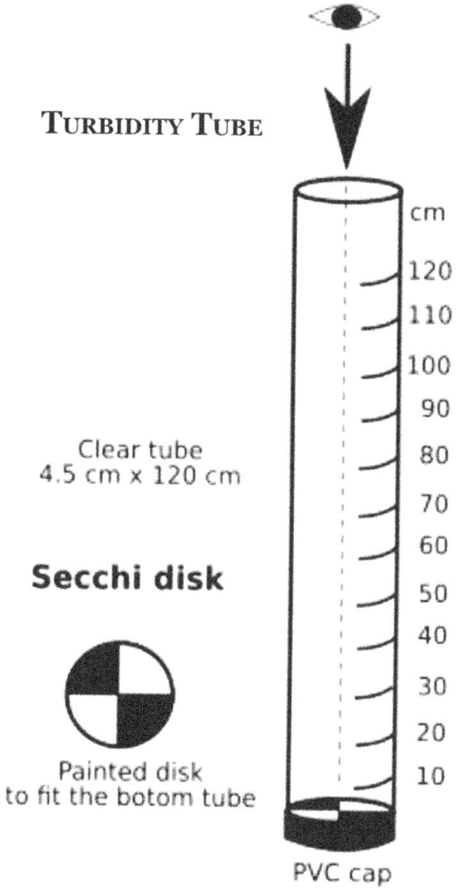


**Length-to-Turbidity Conversion**

| cm | NTU |
|---|---|
| 6.7 | 240 |
| 7.3 | 200 |
| 8.9 | 150 |
| 11.5 | 100 |
| 17.9 | 50 |
| 20.4 | 40 |
| 25.5 | 30 |
| 33.1 | 21 |
| 35.6 | 19 |
| 38.2 | 17 |
| 40.7 | 15 |
| 43.3 | 14 |
| 45.8 | 13 |
| 48.3 | 12 |
| 59.9 | 11 |
| 53.4 | 10 |
| 85.4 | 5 |

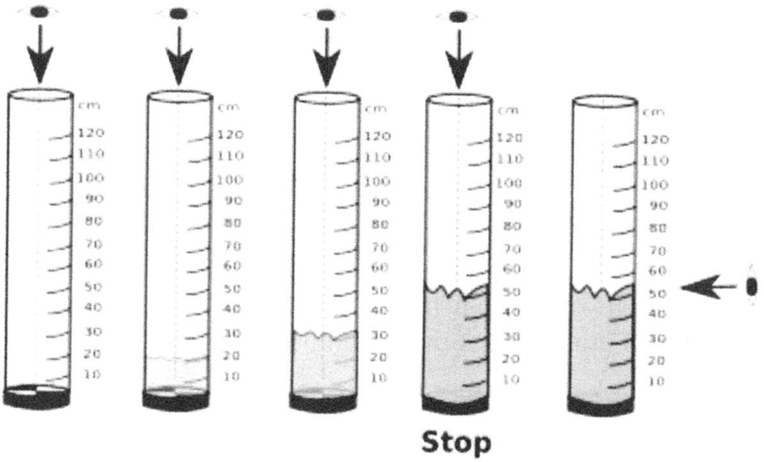


**Fig. 2B.5:** Secchi disk and turbidity measurement.

are using a modified Secchi disk black and white for measuring water turbidity.

Secchi depth readings are thus dependent on lighting conditions and are difficult in shallow systems[16].

### Measuring turbidity procedures

- Take a sample of water from the water source.
- Hold the tube in one hand near the bottom and look into the open end with your head about 10 to 20 centimeters above the tube, so that you can clearly see the Secchi Disk.
- **Adding incremental water sample**. Slowly pour the water sample into the tube, waiting for air bubbles to rise if necessary, until the mark on the bottom of the tube just disappears.
- **Stop adding water sample**. Stop pouring the water sample into the tube and look at the level of water in the tube. For turbidity tubes which have a turbidity scale marked on the side, read the number on the nearest line to the water level. This is the turbidity o f the water.
- **Reading NTU value**. Use the table lenth-to-turbidity conversion to determine the NTU level.
- **Depth in Centimeters** = 244.13 *(Turbidity in NTU)$^{-0.662}$

Homemade turbidity tube[14]

### Homemade turbidity tube

Building a turbidity tube is very easy with materials available at home and easy to get.

- Get a transparent Fluorescent Light Sleeve 1.30 m or > 0.85 cm or a polycarbonated tube for the turbidity tube.
- Use an opaque white plastic milk jug to make the secchi disk: cut a circle that can fit in the bottom in the transparent tube and paint it using a black permanent marker.
- Glue disk onto rubber stopper and insert it into the bottom of the transparent tube.

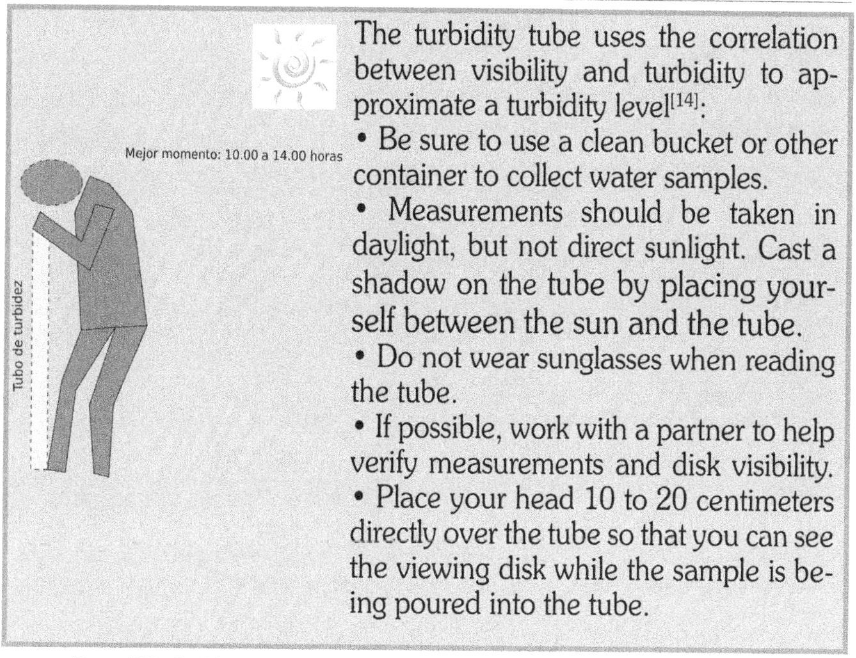


**Fig. 2B.6:** The best time and procedure turbidity measurement.

- Attach a measure tape to the tube and write a small line mark with the NTU value on the tube according to the table lentgh-to-turbidity conversion.
- Trim the top of the tube to about 5 centimeters above the 5 NTU mark.

# References

[1] Luciano de Souza Ferraz, Rener et al. 2017. *Opuntia ficus-indica* (L.) Mill. (Cactaceae) in Climate Change Scenarios and Its Potential for Wastewater Bioremediationin Semi-Arid Regions: A Systematic Review and Meta-Analysis. JEAI, 18(3): 1-11; Article no.JEAI.36730.

[2] Miller, Sarah M. et al. 2008. Toward Understanding the Efficacy and Mechanism of *Opuntia* spp. as a Natural Coagulant for Potential Application in Water Treatment. Environ. Sci. Technol, 42, 4274–4279.

[3] Jayalakshmi , G. et al. 2017. A Review on Native Plant Based Coagulants for Water Purification. International Journal of Applied Environmental Sciences; Volume 12, Number 3 (2017), pp. 469-487.

[4] Barbera , M. and Gurnari, G. 2018. Wastewater Treatment and Reuse in the Food Industry, Chemistry of Foods. Chapter 4: Quality Standards for Recycled Water: *Opuntia ficus-indica* as Sorbent Material.

[5] Lalzad. 2007. An Overview of the Global Water Problems and Solutions. London.

[6] Deshmukh1, Saurabh O. and Hedaoo M. N. 2018. Wastewater Treatment using Bio-Coagulant as Cactus *Opuntia Ficus Indica* – A Review. IJSRD; Vol. 6, Issue 10, 179.

[7] Vijayaraghavan , G. et al. 2011. Application of plant based coagulants for waste water treatment. IJAERS; Vol., Issue, October-December, 2011, 88-92.

[8] Hinrichsen, Don and Tacio, Henrylito. The Coming Freshwater Crisis is Already Here. Population and Water.

[9] Oxford bid ideas geography. Water in the world.

[10] Fox, Dawn I. et al. 2012. Removing Heavy Metals in Water: The Interaction of Cactus Mucilage and Arsenate (As (V)). Environ. Sci. Technol. 2012, 46, 4553–4559.

[11] Young, Kevin Andrew. 2006. The Mucilage of Opuntia Ficus Indica: A Natural, Sustainable, and Viable Water Treatment Technology for Use in Rural Mexico for Reducing Turbidity and Arsenic Contamination in Drinking Water. Kevin Andrew Young.

[12] Lee, Chai Siah et al. 2014. A review on development and application of plant- based bio-flocculants and grafted bio-flocculants. Industrial & Engineering Chemistry Research.

[13] Minesota Pollution Control Agency. 2008. Turbidity: Description, Impact on Water Quality, Sources, Measures - A General Overview. www.pca.state.mn.us.

[14] Myre, Elizabeth & Shaw, Ryan. 2006. The Turbidity Tube: Simple and Accurate Measurement of Turbidity in the Field.

[15] Wikipedia. Secchi disk. https://en.wikipedia.org/wiki/Secchi_disk.

[16] Borok, Aron. 2014. Water Quality Standards. Turbidity Technical Review. State of Oregon. DEQ.

# CHAPTER 3

# NUTRITIONAL PROPERTIES AND FOOD PROCESSING OF "THE TREE OF LIFE AND THE GREEN GOLD"

# NUTRITIONAL PROPERTIES AND FOOD PROCESSING OF "THE TREE OF LIFE AND THE GREEN GOLD"

The hole prickly pear tree or "The Tree of Life and The Green Gold" is edible. Each part of that tree has important nutritional properties and it is considered as a functional food with nutraceutical properties. Cactus pads called nopales or cladodes and fruits called tuna or higo de pala are included in the human diet and also are used as forage. Nopal is a part of the traditional food in the Mexican diet and they were used in human nutrition since at least 9000 years ago.

## INTRODUCTION TO NUTRITIONAL PROPERTIES

*Opuntia* spp. —"The Tree of Life and The Green Gold", an effective eco-friendly natural water biopurification— has been used as human food since at least 9000 years ago or even as early as 12,000 years ago; the prickly pear tree have a high nutritional value, mainly due to their mineral, protein, dietary fiber, and phytochemical compounds; also, it is well known that *Opuntia* cladodes are a good source of dietary fibers, which may help in reducing body weight by binding to dietary fat and increasing its excretion[1]. Cacti have long been considered an important nutritional source and is considered the *bread of the poor in Latin America*[2].

In addition, other studies also demonstrated that nopal possesses a higher content of calcium ($Ca^{2+}$) than other vegetables included in the diet; calcium is consider an important mineral component for bone metabolism; also, the nopal is an important source of nutritional elements such as pectin and mucilage in addition to minerals[3], vitamins and a number of phytochemicals.

*Opuntia* spp. revealed as regards to its nutritional value, high moisture content (about 90%), high in vitro digestibility (about 75%) and a high content of vitamin A; also, showed 29 $\mu$g of carotenoids and 13 mg of ascorbic acid per 100 g of cladodes, organic matter (67%), energy (2.61 Mcal.kg ), crude fiber (4.3%), phosphorus (0.08 - 0.18%), calcium (4.2%), potassium (2.3%) and magnesium (1.4%); a lower protein content was detected (about 5%); the values of organic compounds were expressed in a fresh weight basis and the values of mineral composition were expressed in a dried weight basis[2].

According to M.I. Hernández-Urbiola et al., dietary fiber as well mineral ash composition on nopal increased from 40 days to 135 days[3]. Results for calcium content in nopal powders as a function of the maturity, indicate that the content of this mineral increases with the age of the nopal, from 17.4 mg/g dry weight at 40 days to 34.4 mg/g at 135 days showing an increase of 97%; so, the accumulation of calcium shows that there is a positive relationship with the time of maturity[4].

Studies on the nutritional composition of *Opuntia ficus-indica* (L.) revealed that nutrient composition varies with age, postharvest management and climatic season[11].

### Composition and nutritional value of fruits

The fruits of *Opuntia* spp. contain high concentrations of amino acids, especially proline and taurine, and are rich in minerals such as magnesium (Mg), calcium (Ca), iron (Fe), potassium (K), sodium (Na) and phosphorus (P)[10].

Regardless of the origin or variety of the plant, the fruits are a good natural reservoir of minerals such as magnesium, calcium and potassium and antioxidant components such as vitamin C so they can be used as excellent food supplements[9]. The tuna

| **Nutritional Facts of Cladodes, Raw** | | | |
|---|---|---|---|
| **Nutrient** | **Unit** | **Value/100g** | **%DV** |
| Water | g | 94.12 | |
| Energy | Kcal | 16 | 0.8 |
| Protein | g | 1.32 | 2.64 |
| Total lipid (fat) | g | 0.09 | 0.14 |
| Carbohydrate, by difference | g | 3.33 | 1.2 |
| Fiber, total dietary | g | 2.2 | 7.86 |
| Sugars, total | g | 1.15 | |
| **Minerals** | | | |
| Calcium, Ca | mg | 164 | 12.62 |
| Iron, Fe | mg | 0.59 | 3.28 |
| Magnesium, Mg | mg | 52 | 12.38 |
| Phophorus, P | mg | 16 | |
| Potassium, K | mg | 257 | 7.34 |
| Sodium, Na | mg | 21 | 0.86 |
| Zinc, Zc | mg | 0.25 | 1.28 |
| **Vitamins** | | | |
| Vitamin C, total ascorbic acid | mg | 9.3 | 10.33 |
| Thiamin | mg | 0.012 | 1.2 |
| Riboflavin | mg | 0.041 | 3.15 |
| Niacin | mg | 0.41 | 2.56 |
| Vitamin B6 | mg | 0.070 | 4.12 |
| Fotale, DFE | ug | 3 | 0.75 |
| Vitamin A, RAE | ug | 23 | 2.56 |
| Vitamin A, IU | IU | 457 | |
| Vitamin K, philloquinone | ug | 5.3 | 4.42 |
| **Lipids** | | | |
| Fatty acids, total saturated | g | 0.016 | 0.08 |
| Fatty acids, total monounsatirated | g | 0.018 | |
| Fatty acids, total polyunsaturated | g | 0.050 | |

**Table 3.1:** Cladodes nutrition facts. %DV (% Daily Value) based on 2000 Kcal.

**Source**: USDA Database. https://ndb.nal.usda.gov/ndb/foods/

| Nutritional Facts of Prickly Pear (Fruit), Raw | | | |
|---|---|---|---|
| **Nutrient** | **Unit** | **Value/100g** | **%DV** |
| Water | g | 87.55 | |
| Energy | Kcal | 41 | 2.05 |
| Protein | g | 0.73 | 1.46 |
| Total lipid (fat) | g | 0.51 | 0.65 |
| Carbohydrate, by difference | g | 9.57 | 3.48 |
| Fiber, total dietary | g | 3.6 | 12.86 |
| Sugars, total | g | -- | |
| **Minerals** | | | |
| Calcium, Ca | mg | 56 | 4.31 |
| Iron, Fe | mg | 0.30 | 1.67 |
| Magnesium, Mg | mg | 85 | 20.24 |
| Phophorus, P | mg | 24 | 1.92 |
| Potassium, K | mg | 220 | 4.68 |
| Sodium, Na | mg | 5 | 0.22 |
| Zinc, Zc | mg | 0.12 | 1.09 |
| **Vitamins** | | | |
| Vitamin C, total ascorbic acid | mg | 14 | 15.56 |
| Thiamin | mg | 0.014 | 1.17 |
| Riboflavin | mg | 0.06 | 4.62 |
| Niacin | mg | 0.46 | 2.88 |
| Vitamin B6 | mg | 0.060 | 3.53 |
| Fotale, DFE | ug | 6 | 1.5 |
| Vitamin A, RAE | ug | 2 | 0.22 |
| Vitamin A, IU | IU | 43 | |
| Vitamin K, philloquinone | ug | -- | |
| **Lipids** | | | |
| Fatty acids, total saturated | g | 0.067 | 0.34 |
| Fatty acids, total monounsatirated | g | 0.075 | |
| Fatty acids, total polyunsaturated | g | 0.213 | |

**Table 3.2:** Prickly pears fruit nutrition facts. %DV (% Daily Value) based on 2000 Kcal.

**Source**: USDA Database. https://ndb.nal.usda.gov/ndb/foods/

| Sterols and fat-soluble vitamin profile (g/kg) of cactus pear (*Opuntia ficus-indica* L.) seed and pulp oils[19] | | |
|---|---|---|
| **Compound** | **Seed oil** | **Pulp oil** |
| **Sterols** (g/kg) | | |
| Cholesterol | ND | ND |
| Ergosterol | ND | ND |
| Campesterol | 1.66 | 8.74 |
| Stigmasterol | 0.30 | 0.73 |
| Lanosterol | 0.28 | 0.76 |
| β-Sitosterol | 6.75 | 11.20 |
| Δ-5-Avenasterol | 0.29 | 1.43 |
| Δ-7-Avenasterol | 0.05 | ND |
| **Fat-soluble vitamins** (mg/kg) | | |
| α-Tocopherol | 56 | 849 |
| β-Tocopherol | 12 | 126 |
| γ-Tocopherol | 330 | 79 |
| δ-Tocopherol | 5 | 4220 |
| Total vitamin E | 403 | 5274 |
| β-Carotene | 47 | 420 |
| Vitamin $K_1$ | 525 | 532 |

**Table 3.3:** Sterols and fat soluble compounds present in *Opuntia ficus-indica*. **ND**: No Data.

or prickly pear has value of calcium up to 59 mg/100 g and magnesium up to 98.4 mg/100 g[11]. Fruit juice contains per 100 grams of fresh fruit an amount of fat-soluble vitamins, such as 17.54 μg of β-carotene and 125 μg of vitamin E representing 1.9% to 2.5% and 0.83% of the recommended dietary intake, respectively; it also contains appreciable amounts of vitamins $B_1$ and $B_6$ representing approximately 29% and 42% of the recommended daily dietary intake, respectively; it also covers 73% of vitamin $B_2$ and more than 100% of niacin corresponding to the values of the recommended daily dietary intake[14].

The prickly pear has a composition of 77.2% of carbohydrates, it was confirmed by optical rotation analysis that tuna is composed of 57.1% glucose and 42.9% fructose; these results indicate that cactus fruit can be used as raw material for the manufacture of natural sweeteners; other studies should investigate the use of carbohydrate components of tuna to replace sucrose and other sweeteners used in different food products[20].

Based on the low acidity, the high sugar content and the attractive stable colour of *Opuntia* fruits, they may be very suitable for use as a natural additive or substitute material in the production of many food products[5], such as substitutes for sweeteners and food colourants.

### Composition and nutritional value of cladodes

The nutritional value of cladodes is mainly due to some of their phytochemical components such as ascorbic acid, vitamin E, carotenoids, dietary fibres and amino acids as well as being rich in antioxidant components such as phenolic phytocompounds, flavonoids, tannins, betaxanthin and betacyanin[12] whose properties are considered in the next chapter of this book.

### Composition and nutritional value of mucilage

Mucilage is heteropolysaccharide components that have the ability to absorb water; it is a rich source of carbohydrates and essential minerals, it has good potential for use as a nutritional supplement; mucilage is primarily composed of galactose, mannose, xylose, and other sugars, and, thus, has a high capacity to bind or retain water, similarly to pectins, gums, and some algal polysaccharides; it is an excellent source of essential minerals such as calcium, phosphorus, magnesium, iron, potassium, zinc, and copper; calcium is the most abundant mineral in the mucilage, which agrees with other studies of cactus pads; for example, *Opuntia ficus-indica* was a better source of calcium than spinach, soy, and grains, suggesting that *Opuntia* spp. may have potential actions in the prevention and treatment of diseases such as osteoporosis[23].

Dehydrated and powdered cladodes are a source of dietary fibre that can be used as a natural ingredient in different foods to improve their physical-chemical properties[8].

## "THE TREE OF LIFE AND THE GREEN GOLD" AS A SOURCE OF FOOD

In general, nopal is a traditional food in the Mexican diet —annual per capita consumption of 6.4 kg in 2015— often eaten as a vegetable in salads, pickles, juices, sauces, jams and as a complement in different dishes of the Mexican gastronomy; dehydrated and ground nopal are also ingredients in the food industry to elaborate tortillas, snacks, breads and tortilla chips, among others[8].

The prickly pears are reported to be consumed fresh, after desiccation in sun, in marmalades or used as a colouring agent while "nopals" are consumed in Mexican regions as a constituent of salads[13].

The *Opuntia* fruits can be eaten raw or made into an array of dehydrated, frozen, canned, candy, jam, puree, marmalade, salad dressings, syrup, sauce, pies, fruit leather, health drinks; the pears are often made into smoothie and shakes blending with banana, acai berry or pomegranate; sometimes, exotic cocktail beverage margarita is concocted from the pulp; surplus pears are dried for snacks or dessert toppings[17].

Also is a source of a variety of agri-foods available in more than 50 products such as marmalades, juices, nectars, candies, frozen pulp, alcoholic beverages, pickles, sauces, shampoos, soaps and lotions[25].

Both cladodes and cactus pear fruit are consumed as fresh vegetables, added to casseroles, cooked, canned, or used in salads, syrups, alcoholic drinks, fruit juices and in cheese production; juice obtained from the strained pulp is suggested to be a good source for natural sweeteners and colorants[19].

Some options that offer added value to cactus pear cladodes are their use for the production of flours, soups, jams, dehydrated sheets and desserts; recently, the drying of cactus pear cladodes for the production of flours has been proposed as a viable nutritional alternative[9].

Also, the use of various combinations of yellow betaxanthins and red-purple betacyanins may allow the development of food products without artificial colorants; the organic colorant was incorporated in yoghurt and soft drink much to the acceptance of consumers; the results indicated that the studied *Opuntia* pear cultivars can be developed into food-grade colorants, substituting the synthetic dyes; also betalains can be applied in foods of low acidity without affecting the color[17].

The attraction of consumer, especially children consumers, to red-colored foods such as dairy products, juices, ice-creams, marmalades, confectionery, etc., presents an opportunity not only to use a natural and bioactive colorant to replace synthetic dyes, but also to increase the use of arid and semiarid zones for *Opuntia* crops and consequently improve the quality of life for low-income inhabitants of these areas; betalains could be used as functional ingredients in the food industry, especially for the milky and soft drinks industry[18], without discounting its use in the preparation of home-cooked meals.

Mucilage from cladodes could be used to improve the texture of food products[23] and nopal powders can be an economic alternative when used as a dietary supplement in all seasons, without the need for fresh nopal[24].

## INTRODUCTION TO FOOD PROCESSING INCLUDING NOPAL

Food preparation and cooking influence the nutritional qualities of foods, and potentially their health benefits; these processes can have beneficial effects, for example by improving the digestibility and bioavailability of nutrients, and by enhancing attractiveness to the consumer because of improved texture and taste, and also deleterious effects due to loss of nutrients or the formation of toxic compounds[27].

Most vegetables are commonly cooked before being consumed; it is known that cooking induces significant changes in chemical composition, influencing the concentration and bioavailability of bioactive compounds in vegetables; however, both positive and negative effects have been reported depending upon differences in process conditions and morphological and nutritional characteristics of vegetable species; changes in texture are often dramatic because of the membrane disruption and the associated loss of turgor; in addition, cooked vegetables exhibit poor color quality in comparison with fresh ones[28].

During processing many foods, such as beverages, are subjected to high temperatures to extend their shelf-life and preserve microbiological quality[18], but thermolabile phytochemical compounds can be seriously affected by the temperature.

Also in some vegetables including the prickly pear tree, it can occur physical damage and/or wounds caused by handling increase the rate of respiration and the production of ethylene gas quickly results, with associated increases in the rates of other biochemical reactions responsible for changes in colour (including browning), flavour, texture and nutritional value[22].

For human consumption, cactus cladodes are usually eaten peeled, fresh or cooked by boiling, grilling, frying, or also as juices and sauces[15]. Fruits have been used for the preparation of several food products such as juices, marmalades, etc.[10]. So, when "The Tree of Life and The Green Gold" is prepared and processed applying any cooking technique, it will suffer physical and biochemical reactions and changes in the final product proper of the cooking method applied.

Before to start using in the kitchen any part of the prickly pear tree, removing thorns and cleaning are the first processing procedures that need to be done, trimming and chopping are also part of the main cooking techniques applied.

## EFFECT OF THERMAL TREATMENTS ON *OPUNTIA*

Traditionally, thermal treatment is used to extend *Opuntia* beverages shelf-life; however, heat processing may lead to degradation of thermolabile compounds and change the organoleptic properties of the product; for this reason, the application of innovative non-thermal processing technologies such as high pressure processing, high pressure homogenization, pulsed electric fields and ultrasounds can constitute useful tools in order to retain beneficial thermolabile compounds in the food products[10].

Heat treatment can induce changes in chemical and nutritional qualities of foods such as water loss, changes in energy, fat content and consequently in the fatty acid profile, as well as in the amount and bioavailability of bioactive compounds such as (poly)phenols; cooking techniques produce structural alterations in cactus cladodes which induce changes in the nutritional composition, their (poly)phenolic compounds and consequently their antioxidant capacity, depending on the technological conditions[15].

Also, thermal treatments can damage nutritive value and natural texture, taste, flavor, and color in foods; some of the most important bioactive compounds in foods such as vitamins, polyphenols, and other compounds are thermally instable; consequently, optimal exposure to high temperatures provides a way

| Phenolic compounds (mg./100 g dry weight basis)[16] | | | | | |
|---|---|---|---|---|---|
| Phenolic compound | Fresh | Steaming | Microwaving | Boiling | Grilling |
| Gallic acid | 11.35 | 6.75 | 12.34 | 5.30 | 7.81 |
| Vanillic acid | 9.96 | 4.31 | 0.97 | 2.96 | ND |
| 4-Hydroxybenzoic acid | 10.42 | 3.96 | 4.05 | 1.12 | 4.94 |
| Syringic acid | 3.39 | 1.42 | ND | 2.04 | 1.72 |
| Epicatechin | 4.65 | 2.82 | 1.87 | 3.36 | 2.03 |
| Catechin | 15.81 | 17.76 | 13.40 | 5.15 | 3.12 |
| Ferulic acid | 7.00 | 7.03 | 6.72 | 1.34 | 6.50 |
| Rutin | 5.36 | 4.96 | 5.62 | ND | 4.58 |
| Vanillin | 7.87 | 5.63 | 7.84 | 2.62 | ND |
| Protocatechuic acid | 26.56 | 4.10 | 4.92 | 3.22 | 3.59 |
| Quercitrin | 17.3 | 0.74 | 1.15 | 3.22 | 0.84 |
| Quercetin | 5.19 | 7.61 | 9.33 | 5.82 | 7.55 |
| Kaempferol | 13.90 | 10.01 | 13.69 | 9.04 | ND |

**Table 3.4:** Variation of phenolic compounds in *Opuntia ficus-indica* according to the influence of the cooking method.

to use sufficient amount of heat for safe manufacturing practices, but short enough to save natural food quality[10].

Some of the most important bioactive compounds found in foods such as vitamins, polyphenols and other phytochemicals are usually thermally unstable; also, natural components such as colorants, antioxidants and other phytocomponents extracted from the fruits of *Opuntia* spp. are sensitive to heat treatments when the temperatures applied are between 50°C and 90°C; heat treatment was found to reduce vitamin C amounts from 46% to 76%, reduce total phenols from 27% to 52%, flavonoids to 52%, betalinates from 7% to 45%, and also reduces antioxidant capacity from 16% to 45%[10].

Temperature can be considered the most important factor affecting the betalains stability, both in its processing and in its storage; betalains are labile compounds, and can be lost when high temperatures are used[18].

As an illustrative example, the table 3.4 shows how affect different cooking methods on several phenolic compounds that are present in *Opuntia* spp. According to the results, cooking *Opuntia* spp. by steaming and microwave heating is recommended to ensure a higher retention of the phenolic compounds and antioxidant constituents, and these methods should be selected to preserve the nutraceutical qualities of the food[16].

The high temperatures and the long drying time can result in a loss of nutritional value, color, taste and of functional properties of foods; the application of high thermal treatments during the preparation of cactus pear cladodes flours affect the water absorption capacity (functional property) and the green color negatively; presenting higher functional qualities, the cladodes flours processed at 60°C are considered the most adequate for the formulation of dietary supplements[9].

## "THE TREE OF LIFE AND THE GREEN GOLD" FOOD PREPARATION AND RECIPES

After removing thorns and cleaning with fresh water the parts of the prickly pear tree, the raw material will be ready to prepare any kind of foods. To avoid nutrient losses it is a good idea to think in preparing plates with raw nopal parts. Also, it is a good idea trim and chop cladodes that will be eaten raw few minutes before eating them because when cells are damaged occurs enzymatic reactions that modify the properties of the food.

When cooking cladodes, do not overcook to avoid loss of thermosensitive nutrients.

Because sugar is not good for health, especially for people who have diabetes, it is advisable to avoid adding sugar in the preparation of cactus-based sweets such as jams.

Since sugar is not an essential nutrient, it has no recommended daily allowance, or RDA. Based on the is increasing concern that intake of free sugars, particularly in the form of sugar-sweetened beverages, WHO recommends a reduced intake of free sugars throughout the life course, recommends reducing the intake of free sugars to less than 10% of total energy intake and also suggests a further reduction of the intake of free sugars to below 5% of total energy intake. Following the WHO recommendation it is a good idea do not add sugar for preparing food such as marmalades and avoid preparing those that are requiring added sugar.

### Removing thorns

Both cladodes and fruits have thorns that need to be removed. To remove thorns from cladodes and fruits different techniques are applied.

- From cladodes, thorns can be removed using a sharp knife; just carefully slice the thorny nods from the nopal paddle and trim around the outside edge.

- For removing thorns from prickly pear fruits, just a quick pass over a fire to remove the glochids.

### Cladodes preparation

To prepare the pad, simply hold its base and scrape the skin on both sides with a blunt knife until all the spines are removed; then peel the pads and cut them into shoestring strips or dice them according to the needs of the recipe; they can be eaten raw in salads, boiled and fried like eggplant, pickled with spices, or cooked with shellfish, pork, chilies, tomatoes, eggs, coriander, garlic, and onions; recipes range from appetizers, soups, and salads through entrees, vegetable dishes, and breads to desserts, beverages, and candies[26]

Never overcook and select the best cooking method for preparing hot dishes.

#### Azteca salad

Azteca salad is prepared using nopalitos or young cladodes. The salad contains tuna fish, beans, diced cladodes and chillies or hot jalapeño peppers.

Combine the ingredients in quantities as you like and serve in the plate with other foods such as rice.

#### Healthy sauces

Sauces are made using ground or chopped tender cladodes, with the addition of various kind of vegetables such as peppers, tomatoes, onions and spices in varying proportions.

Use a casserole or a saucepan with lid.

Put all the ingredients chopped with spices at your choice.

**Do not add water**. Cladodes and vegetables have about 70-90% of water. With the heat cell suffer damages and they release the water.

Simmer for 10-15 minutes.

Applying these procedures you will obtain a nutritious and delicious sauce.

The sauce can be used for seasoning several kind of dishes such as brown rice, integral pasta, several types of meat, fish and shellfish.

### Prickly pear juices and smoothies

Blends the prickly pear fruit with other fruit juices such as pineapple because mixed fruits could enhance the quality of the juices; pineapple juice and the addition of citric acid decreased the pH of the blends and increased their microbiological stability[26].

### Tuna marmalade

After to remove and discards both ends of the tunas that you like to made marmalade and dice the fruits.

Put the diced tunas in a clean glass jar with lid that allow evaporation. Cook using Bain-Marie until a consistent paste is obtained. Put the consistent paste in sterilized pot glass, as an antimicrobial agent you can put in the top a thin film of cinnamon, put the lid and leave in Bain-Marie for 20 minutes.

The marmalade will be enough sweet with a great natural tuna flavor.

| Tuna pineapple orange smoothie | |
|---|---|
| **Ingredients** | **Preparation** |
| 2 prickly pears<br>1 orange<br>200 g pineapple | Remove and discards both ends of the tunas.<br>Cut in slice.<br>Peel the orange.<br>Cut the orange in pieces.<br>Cut pineapple in small pieces.<br>Put all in the blender and blend them.<br>Serve in a glass. |

## TOXICOLOGY

Orally, *Opuntia ficus-indica* is usually well tolerated. However, it has been reported that it may cause mild diarrhea, nausea, increased stool volume, increased stool frequency, abdominal fullness, and headache and low colonic obstruction[6].

Cactus pear extracts can be easily used as dietary supplements as it has no toxic effects[13].

# References

[1] Santos Díaz, María del Socorro et al. 2017. Opuntia spp.: Characterization and Benefits in Chronic Diseases. Oxidative Medicine and Cellular Longevity.

[2] Chauhan, S. P. et al. 2010. Biological Actions of Opuntia Species. Systematic Reviews in Pharmacy, Vol 1 | Issue 2.

[3] Hernández-Urbiola, M.I. et al. 2010. Study of Nutritional Composition of Nopal (Opuntia ficus indica cv. Redonda) at Different Maturity Stages. The Open Nutrition Journal, 2010, 4, 11-16.

[4] Contreras-Padilla, Margarita et al. 2011. Evaluation of oxalates and calcium in nopal pads (Opuntia ficus-indica var. redonda) at different maturity stages. Journal of Food Composition and Analysis 24 38–43.

[5] Chiteva, Rose and Norman Wairagu. 2013. Chemical and nutritional content of Opuntia ficus-indica (L.). Vol. 12(21), pp. 3309-3312.

[6] Ondarza, Mauricio A. 2016. Cactus Mucilages: Nutritional, Health Benefits and Clinical Trials. J. Med. Bio. Sci. Res.

[7] Toit, Alba du et al. 2018. Cultivar and Harvest Month Influence the Nutrient Content of Opuntia spp. Cactus Pear Cladode Mucilage Extracts. Molecules, 23, 916.

[8] Ciriminna, Rosaria et al. The unfolding bioeconomy of nopal (*Opuntia* spp.).

[9] López-Cervantes, Jaime et al. 2011. Functional properties and proximate composition of cactus pear cladodes flours. Ciênc. Tecnol. Aliment., Campinas, 31(3): 654-659.

[10] Barba, Francisco J. et al. 2017. Impact of conventional and non-conventional processing on prickly pear (*Opuntia* spp.) and their derived products: From preservation of beverages to valorization of by-products. Trends in Food Science & Technology 67, 260-270.

[11] Stintzing, Florian C. et al. 2001. Phytochemical and nutritional significance of cactus pear. Eur Food Res Technol, 212 : 396–407.

[12] Hfaiedh Mbarka et al. 2018. Phytochemical analysis and nephroprotective effect of Cactus (*Opuntia ficus-indica*) cladodes on sodium dichromate-induced kidney injury in rats. Appl. Physiol. Nutr. Metab.

[13] Pavithra K. et al. 2015. Pectin and Isolated Betalains from *Opuntia dillenii* (Ker-Gawl) Haw. Fruit Exerts Antiproliferative Activity by DNA Damage Induced Apoptosis. IJPPR, Volume 7, Issue 6.

[14] Hassan Abd El-Razek, Fatma and Hassan, Amal A. 2011. Nutritional Value and Hypoglycemic Effect of Prickly Cactus Pear (Opuntia Ficus-Indica) Fruit Juice in Alloxan-Induced Diabetic Rats. Australian Journal of Basic and Applied Sciences, 5(10): 356-377.

[15] De Santiago, Elsy et al. 2017. Impact of cooking process on nutritional composition and antioxidants of cactus cladodes (*Opuntia ficus-indica*). Food Chemistry.

[16] Cortez-García, Rosa María et al. 2015. Effects of Cooking Methods on Phenolic Compounds in Xoconostle (*Opuntia joconostle*). Plant Foods Hum Nutr.

[17] Patel, Seema. 2012. Reviewing the prospects of Opuntia pears as low cost functional foods. Springer Science+Business Media B.V.

[18] Sáenz, Carmen et al. 2012. Red betalains from Opuntia spp.: Natural colorants with potential applications in foods. Israel Journal of Plant Sciences, Vol. 60 pp. 291–299.

[19] Ramadan, Mohamed Fawzy and Mörsel, Jörg-Thomas. 2003. Oil cactus pear (*Opuntia ficus-indica* L.). Food Chemistry 82, 339–345.

[20] Sáenz, Carmen et al. 1998. Cactus pear fruit: A new source for a natural sweetener. Plant Foods for Human Nutrition 52: 141–149.

[21] Cota-Sánchez, J. Hugo. 2016. Chapter 28: Nutritional Composition of the Prickly Pear (*Opuntia ficus-indica*) Fruit. Elsevier.

[22] Sáenz, Carmen et al. 2013. Agro-industrial utilization of cactus pear. Food and Agriculture Organization of the United Nations, FAO.

[23] Monrroy, Mariel et al. 2017. Extraction and Physicochemical Characterization of Mucilage from *Opuntia cochenillifera* (L.) Miller. Journal of Chemistry.

[24] Hernández-Urbiola, Margarita I. 2011. Chemical Analysis of Nutritional Content of Prickly Pads (*Opuntia ficus indica*) at Varied Ages in an Organic Harvest. Int. J. Environ. Res. Public Health, 8.

[25] Kauthale, Vitthal et al. 2017. Cactus an Emerging Fodder Crop of Arid and semi Arid India. BAIF Development Research Foundation.

[26] Shedbalkar, Utkarsha U. et al. 2010. Opuntia and Other Cacti: Applications and Biotechnological Insights. Tropical Plant Biol. 3:136–150.

[26] University of California. 1989. Prickly Pear Cactus Production. Small Farm Center, University of California.

[27] Hoffman, Richard and Gerber, Mariette. 2015. Food Processing and the Mediterranean Diet. Nutrients 2015, 7.

[28] Miglio, Cristiana et al. 2008. Effects of Different Cooking Methods on Nutritional and Physicochemical Characteristics of Selected Vegetables. J. Agric. Food Chem. 56, 139–147.

# CHAPTER 4

# FUNCTIONAL AND NUTRACEUTICAL PROPERTIES AND PHYTOCHEMICAL COMPOSITION OF "THE TREE OF LIFE AND THE GREEN GOLD"

# Functional and Nutraceutical Properties and Phytochemical Composition of "The Tree of Life and The Green Gold"

"The Tree of Life and The Green Gold" or prickly pear is a plant that contains important bioactive compounds that possess properties that benefit health. It is a functional food with nutraceutical properties. The prickly pear has been used in traditional folk medicine to effectively treat various types of health conditions, in particular is widely used by Mexicans. *Opuntia* spp. is a source of important bioactive compounds that have a wide range of biological functions. Compounds extracted from cladodes, flowers and fruits have been shown to have a number of pharmacological actions that greatly benefit health, in addition to being an effective natural biopurifier of water.

## INTRODUCTION

The prickly pear, *Opuntia* spp., is a plant that in addition to containing important phytochemical compounds with nutraceutical properties, can grow not only in arid and semi-arid areas but can allow and sustain the development of life in adverse areas. Due to its surprising characteristics, it deserves to be called "The Tree of Life and Green Gold".

The cladodes of prickly pear tree contain bioactive compounds, well-known for their health-related properties[1]. It has been used in traditional folk medicine because of its role in treating a number of diseases and conditions, including anti-inflammatory effects, hypoglycemic effects, inhibition of stomach ulceration, neuroprotective effects; through antioxidant actions and also used for treating diabetes, burns, bronchial, asthma and indigestion in many countries over the world[3]. Folk medicine, particularly at Mexican population, records the healing of a large number of diseases with the use of this plant[2].

The *Opuntia* cladodes have high content of fiber and beneficial effects in the metabolism of glucose; in addition, minerals, especially potassium and calcium, as well as phenol contents are present in the cladodes[2]. Interestingly, alkaloids, indicaxanthin, neobetanin, and various flavonoids have been isolated from the cactus[6].

## PHYTOCHEMICAL COMPOUNDS AND THEIR FUNCTIONAL PROPERTIES

The phytochemical composition on dry weight of commercial and wild species of *Opuntias* spp. depends on variety and species, maturation stage, and environmental conditions[4] and the part of the plant as show table 4.1.

The cactus plant can be exploited in total because its bioactive components can be extracted from different parts of its anatomy: flowers, fruit, cladodes, roots and seeds[21] (see table 4.1).

For example, cladodes contain vitamins, antioxidants, and various flavonoids, particularly the highly efficient radical antioxidant 3-methyl ether quercetin; for every 100 grams of dry matter, the flower contains 4900 mg of gallic acid and 4269 mg of 6-isorhamnetin 3-O-robinobioside which are part of its main phyto compounds; for every 100 grams of dry matter from fruit pulp, 218 was obtained.8 mg phenol together with a high content of isorhamnetin glycosides (50.6 mg) compared to other flavonoids; the fruit, in particular its skin, may contain up to 17.6 g/kg of vitamin E corresponding to $\alpha$-tocopherol[6].

### PHYTOCHEMICAL COMPOSITION OF FRUITS

The fruits of *Opuntia* spp. have become popular due to their nutritive value and health benefits, including antioxidant, antiath-

| Phytochemical composition of *Opuntia* spp.[6] | | | | | |
|---|---|---|---|---|---|
| **Main component identified** | **Flower** | **Pulp** | **Seed** | **Skin fruits** | **Cladodes** |
| Gallic acid | 1630-4900 | | | | 0.64 - 2.37 |
| Quercetin 3-O-Rutinoside | 709 | | | | |
| Quercetin 3-O-Glucoside | 447 | | | | |
| Isorhamnetin 3-O-Robinobioside | 4269 | | | | |
| Isorhamnetin 3-O-Galactoside | 979 | | | | |
| Isorhamnetin 3-O-Glucoside | 724 | | | | 4.59– 2.21 |
| Kaempferol 3-O-Arabinoside | 324 | | | | |
| Total phenolic acid | | 218.8 | 48 89 | 45700 | |
| Quercitin | | 9 | | 4.32 | |
| Isorhamnetin | | 4.94 | | 2.41-91 | |
| Kaempferol | | 0.78–2.7 | | 0.22 | |
| Luteolin | | 0.84 | | | |
| isorhamnetin glycosides | | 50.6 | | | |
| Feruloyl-sucrose isomer 1 | | | 7.36-17.62 | | |
| Feruloyl-sucrose isomer 2 | | | 2.9–17.1 | | |
| Feruloyl-sucrose isomer 1 | | | 12.6–23.4 | | |
| Total Flavonoid | | | 1.5–2.6 | 6.95 | |
| Total Tannins | | | 4.1–6.6 | | 14.08–16.18 |
| Coumarin | | | | | 0.06–5.02 |
| 3,4-dihydroxybenzoic | | | | | 0.5–4.72 |
| 4-hydroxybenzoic | | | | | 0.56–34.77 |
| Ferulic acid | | | | | 0.58–3.54 |
| Salicylic acid | | | | | 2.29–39.67 |
| Isoquercetin | | | | | |
| Nicotiflorin | | | | | 2.89–146.5 |
| Rutin | | | | | 2.36–26.17 |
| Narcissin | | | | | 14.69–137.1 |

**Table 4.1:** Phytochemical composition of the different parts of *Opuntia* spp. Nutraceutical properties vary in different parts of the plant.
**Source**: USDA Database. https://ndb.nal.usda.gov/ndb/foods/

erogenic, and antiulcerogenic properties, with protective effects against low-density lipoproteins peroxidation[15].

The identification and quantification of betalains in the fruits of ten varieties of *Opuntia* —purple, red, orange, yellow and white pulp— were studied by Castellanos-Santiago and Yahia; these authors found a total of 24 known and unknown betalains, including betaxanthins and betacyanins; they also found a higher betalain content in the purple fruits that was of the order of 8.1 mg/g of dryed fruit, while the betacyanin content in the skin of *Opuntia ficus-indica* fruits was approximately 1.4 times higher than in the pulp; they also found that *Opuntia ficus-indica* fruits contained 99% indicaxanthin, a betaxanthin characteristic of *Opuntia*[15].

Cactus fruit contains substantial amounts of ascorbic acid, vitamin E, carotenoids, fibers, amino acids and antioxidant compounds such as phenols, flavonoids, betaxanthin and betacyanin; *Opuntia ficus indica* flowers contains different flavonoids notably kaempferol and quercetin[6].

Because cacti are genetically very diverse, their fruits are a promising source of natural food colouring; due to the structural diversity of the red-violaceous betacyanins and their yellow-orange betaxanthins found in the edible matrix of the fruit, it is possible to obtain a wide range of colour hues; the highest levels of betalains in the flesh and shell of *Opuntia ficus-indica* amounted to 113.9 mg/100 g of fresh fruit[17]. The nopal fruit also contains substantial amounts of ascorbic acid, vitamin E, carotenoids, fibres, amino acids and antioxidant compounds such as phenols, flavonoids, betaxanthin and betacyanin while the flowers of *Opuntia ficus-indica* contain different flavonoids, in particular kaempferol and quercetin[6]. Also, the fruits of *Opuntia* spp. are characterized by high levels of amino acids, especially proline and taurine[17].

The tunas, that is to say the fruits of *Opuntia* spp. should be consumed preferably with the peel because they have some important phytochemical compounds with nutraceutical properties. The peel of the *Opuntia ficus-indica* fruit contains mainly

| Cladodes: Amino acids composition | |
|---|---|
| **Amino acids** | **mg/100 g** |
| Tryptophan* | 14 |
| Threonine* | 40 |
| Isoleucine* | 49 |
| Leucine* | 77 |
| Lysine* | 59 |
| Methionine* | 15 |
| Cystine | 8 |
| Phenylalanine* | 49 |
| Tyrosine | 29 |
| Valine* | 59 |
| Arginine | 52 |
| Histidine* | 25 |
| Alanine | 50 |
| Aspartic acid | 86 |
| Glutamic acid | 145 |
| Glycine | 46 |
| Proline | 43 |
| Serine | 43 |
| *Essential amino acids. | |

**Table 4.2:** Amino acid composition of cladodes.

| Cladodes: Lipids composition | |
|---|---|
| **Lipids (Fat)** | **mg/100 g** |
| Fatty acids, total saturated | 16 |
| 16:0 - Palmitic acid | 12 |
| 18:0 - Stearic acid | 2 |
| Fatty acids, total monounsaturated | 18 |
| 16:1 undifferentiated | 2 |
| 18:1 undifferentiated | 17 |
| Fatty acids, total polyunsaturated | 50 |
| 18:2 undifferentiated | 44 |
| 18:3 undifferentiated | 5 |
| Cholesterol | 0 |
| Fatty acids, total trans | 0 |
| Total lipids (fat) | 90 |

**Table 4.3:** Fat composition of cladodes.

| Cladodes: Sugar composition[1] | |
|---|---|
| **Monosaccharides** | **ug/mg** |
| Rhamnose | 7.13 |
| Arabinose | 39.64 |
| Xylose | 18.64 |
| Mannose | 13.64 |
| Galactose | 33.69 |
| Glucose | 153.15 |
| Galacturonic acid | 96.26 |

**Source**: USDA Food database.

**Table 4.4:** Sugar composition of cladodes.

| Detectable amino acids content in the fruit pulp[21] | | | | | |
|---|---|---|---|---|---|
| Amino acid | mg/L | Essential Amino acid | mg/L | Amino acid | mg/L |
| Alanine | 87.2 | Histidine | 45.2 | Taurine | 434.3 |
| Arginine | 30.5 | Isoleucine | 31.2 | Citrulline | 16.3 |
| Asparagine | 41.7 | Leucine | 20.6 | | |
| Glutamic acid | 66.1 | Lysine | 17.4 | | |
| Glutamine | 346.2 | Methionine | 55.2 | | |
| Glycine | 11.3 | Phenylalanine | 23.3 | | |
| Proline | 1265.2 | Threonine | 12.6 | | |
| Serine | 174.5 | Tryptophan | 12.3 | | |
| Tyrosine | 12.3 | Valine | 39.4 | | |

**Table 4.5:** Amino acids content present in the fruit pulp of *Opuntia ficus-indica.*

isorhamnetin derivatives of flavonols, such as isorhamnetin-3-O-rutinoside, diglycoside isorhamnetin, triglycoside isorhamnetin and other glycoside isorhamnetin that are present but in small amounts[15].

## Phytochemical composition of mucilage

*Opuntia* genus is widely known for its mucilage production; mucilage, is a complex carbohydrate with a great capacity to absorb water and contains different proportions of L-arabinose, D-galactose, L-rhamnose, and D-xylose, as well as galacturonic acid; in some countries, small farmers use cactus mucilage to purify drinking water; also, the cladodes dehydrated and ground into powder are source of dietary fiber which may be used as natural ingredient in different foods to enhance their beneficial properties[8].

## Phytochemical composition of Cladodes

The composition of the sugars in *Opuntia* spp. was studied from the cladodes; the cladodes are composed of different monosaccharides; each milligram extracted and dried contains 7.13 rhamnose ug, 0.74 fucose ug, 39.64 arabinose ug, 18.64

| *Opuntia ficus-indica* functional and nutraceutical properties[21] | | | |
|---|---|---|---|
| **Part** | **Extraction** | **Action** | **Studied system** |
| Cladodes. | Extracts. | Antiviral. | Intracellular virus replication inhibition and extracellular virus inactivation.<br>Guinea pig. |
| | | Hypoglycemic and antidiabetic effects | Diabetic rats. |
| | | Anti-hyperlipidemic effect and cholesterol reducing action. | Rats. |
| | | Antiobesity factor | Humans. |
| | | Anti-inflammatory actions | Humans. |
| | indicaxanthin | | Pleurisy rats |
| | Flavonoids extracts | Neuroprotection against neuronal oxidative injuries. | Primary cultured rat cortical cells. |
| | | Antiulcerogenic effects and antigastritis | Rats. |
| | Extracts | Protection against nickel-induced toxicity. | Rats. |
| | | Protection against oxidative damage induced by zearalenone | Mice. |
| | | Diuretic effect | Rats. |
| | | Healing properties | Humans. |
| Plant | Extracts | Neuroprotection against cerebral ischemia | In vitro studies in cultured mouse cortical cells<br>In vivo<br>studies in gerbils. |
| | Extracts | Alcohol hangover symptoms alleviation | Humans |
| Seed oil | | Anti-hyperlipidemic effect and cholesterol reducing action | Rats |
| | | Hypoglycemic and antidiabetic effects | Rats |
| Seed | | Anti-hyperlipidemic effect and cholesterol reducing action | Rats |
| | | Hypoglycemic and antidiabetic effects | Rats |
| Flower | | Diuretic effect | |

| *Opuntia ficus-indica* funtional and nutraceutical properties[21] | | | |
|---|---|---|---|
| **Part** | **Extraction** | **Action** | **Studied system** |
| Fruit | Juice | Antiulcerogenic effects and antigastritis | Rats |
| | Extract | Decreasing effect on oxidative stress | Human |
| | | DNA damage reduction | Human peripheral lymphocytes |
| | | Diuretic effect | Rats |
| | Juice and extract | Liver protection | Liver |
| | Aqueous extracts | Cancer preventive properties | Ovarian and cervical epithelial cells, as well as ovarian, cervical, and bladder cancer cells, ovarian cancer cells, leukemia, cell lines |
| | | Improvement of platelet function | Human |

**Table 4.6:** Functional and nutraceutical properties of the different parts of "The Tree of Life and The Green Gold".

xylose ug, 13.64 mannose ug, 33.69 galactose ug, 153.15 glucose ug, and 96.26 galacturonic acid ug[1].

The fatty acids compositions in the lipids from cactus pear cladodes flours identified were the following: acid (C14:1n5), myristic acid (C14:0), palmitoleic acid (C16:1n7), palmitic acid (C16:0), linoleic acid (C18:2n6), linolenic acid (C18:3n3), oleic acid (C18:1n9), vaccenic acid (C18:1n7), stearic acid (C18:0), arachidic acid (C20:0), gondoic acid (C20:1n9) and behenic acid (C22:0); in cactus pear cladodes flours, the most abundant fatty acids were palmitic acid (C16:0), linoleic acid (C18:2n6), linolenic acid (C18:3n3), and oleic acid (C18:1n9), which corresponded to 88%, 91% and 92% of the total fatty acids present in the cactus pear cladodes flours prepared in this study; furthermore, the lipids present in the samples contained about 50% of the essential fatty acids (C18:2n6, C18:3n3), while the saturated fatty acid content was 37%[5].

## PHARMACOLOGICAL PROPERTIES

*Let food be your medicine.*
Hipócrates (460 a.C.-377 a.C.).

In folk medicine, cactus plants have been used by Native Americans for centuries as a dietary supplement; more recently, cladodes, fruits, and flowers have been traditionally used in folk medicines in several countries; cladodes are still used for gastric ulcer treatment and for its healing activity as therapeutic agents; the consumption of cactus fruits and their juices has traditionally been recommended for their diuretic effect, functions as hypoglycaemic agent, anti-allergic, analgesic and anti-inflammatory actions, and for gastritis relief[6].

In traditional medicine, *Opuntia ficus-indica* has been used for the treatment of burns, wounds, edema, hyperlipidemia, obesity and catarrhal gastritis; alcoholic extracts are indicated for anti-inflammatory, hypoglycemic, and antiviral purposes; also, the *Opuntia ficus-indica* var. saboten is used for wounds, burns, edema, dyspepsia, and as a neuroprotective, cytoprotective, antispasmodic, and chemopreventive; the extracts of fruits have shown to have antioxidant, and anti-inflammatory activities; and clinical studies have shown the effect of nopal on type 2 diabetic patients, suggesting that the anti-hyperglycemic effect may be due to its fiber and pectin content, which may decrease carbohydrate absorption; also, it was observed that *Opuntia* extracts have anti-tumoral and hepatoprotective effects in mice[8].

*Opuntia* cactus has been recognized as a functional food and a source of nutraceuticals, a food or part of a food that provides medical or health benefits, including prevention and/or treatment of a disease; the fruit and cladodes of cactus pear provide interesting sources of functional compounds, including fibre, hydrocolloids or mucilage, pigments such as betalains and carotenoids, minerals, such as calcium and potassium and vitamins with antioxidant properties, such as vitamin C[8].

*Opuntia* spp. is a source of novel bioactive compounds that have a wide range of biological functions[4]. Extracted compounds from cladodes have shown a number of pharmacological actions, including emollient and wound-healing effects, hypocholesterolemic effects, inhibition of stomach ulceration, neuroprotective effects through antioxidant actions, and anti-inflammatory effects; also they showed effects on blood glucose control, serum lipid levels in add to antiglycated activities that have been related to the hypoglycaemic and antidiabetic properties[1].

Prickly pear tree contain mucilage —complex polysaccharides with extremely branched structures— and pectin considered soluble dietary fibers that exert physiological effects like hypocholesterolemic and hypoglycemic in humans[8].

The cladodes of *Opuntia* spp. contain a viscous substance called mucilage which is a hydrocolloid with promising functional and rheological properties; hydrocolloids are used to modify the textural properties of foods especially when it is necessary to omit some ingredients such as gluten, fats, eggs and dairy products; hydrocolloids enable the development of low-calorie, allergen-free functional food products; mucilage, as mentioned above, is capable of absorbing and retaining large amounts of water in its structure; the development of functional food products with mucilage would not only benefit consumers by providing foods, medicines and natural treatments to treat chronic diseases such as cancer, diabetes mellitus and cardiovascular disease, but could also promote the development of commercial cactus crops since mucilage has the ability to act as a functional ingredient for the manufacture of a number of commercial food products; cactus mucilage is a promising hydrocolloid that could replace a number of undesirable ingredients in various food products[19].

As already mentioned, *Opuntia* spp. is rich in dietary fibre. The components of dietary fiber are formed by various chemical components such as cellulose, hemicellulose, pectin, lignin and gum among others that are resistant to digestive enzymes; the

**Fig 4.1:** *"Tenochtitlan fundation myth in the stone sculpture Teocalli of the Sacred War, back view. An eagle that is holding the atl tlachinolli (duality symbol) with its beak is resting over an* ***opuntia (nopal)*** *which emerges from the body of Copil (aztec prince), according to aztec mythology".*

**Source**: https://en.wikipedia.org/wiki/File:Tenochtitlan_fundation_myth_in_the_Teocalli_of_the_Sacred_War_stone_sculpture_back_view.svg.

benefits associated with fiber content, as mentioned above, are well known in the prevention of diseases such as diabetes, the treatment of gastrointestinal disorders, diseases associated with low dietary fiber intake, reduction of blood glucose values, and antihyperlipidemic and antihypercholesterolemic effects[24].

Scientific studies showed that *Opuntia ficus-indica* tissues revealed antiviral, anticancer, hypoglycemic, antioxidant and diuretic activity; the fruits have anti-inflammatory and analgesic effects; also, Tesoriere et al. studied the anti-inflammatory activity of indicaxanthin, another pigment from the edible fruit of *Opuntia ficus indica* and showed that the pigment may have the potential to modulate inflammatory processes at the intestinal level; the cladodes showed antiulcer property and Ennouri et al. suggested that the seed oil might have hypocholesterolemic and hypolipidemic activities[2]. In vivo and in vitro experiments demonstrated beneficial health activities of flavonoids as protective agents against cancer, cardiovascular, inflammatory, and allergic disorders, capillary fragility, and human platelet aggregation[1].

In the sub-Saharan traditional medicine pharmacopeia, cactus flowers and fruits are given as anti-ulcerogenic or antidiarrheal agents; flowers being also administered as an oral antihemorrhoid medication and cladode sap as a treatment for whooping cough; cladodes of *Opuntia ficus-indica* extracts may lower cholesterol level and convey antiulcer and anti-inflammatory mechanisms, and the water extract remarkably improves wound healing[6].

The high level of calcium and magnesium are useful in the prevention of osteoporosis and cramps, respectively; whilst high levels of calcium, magnesium and potassium are used for energy and sports drinks to uphold the mineral pool during periods of physical exhaustion, low levels of sodium and chloride are preferred for preventing high blood pressure[17].

The fresh young cladodes are a source of proteins including essential amino acids, and vitamins; several studies have reported that high levels of amino acids especially proline, taurine and serine are found in nopal[7].

Taurine, a non standard amino acid of plant origin, which was formerly of little interest, has recently been reevaluated in nutritional science; it is already added to sports and energy drinks and claimed as a major ingredient in alcoholic drinks with a soft drink flavour; taurine has specific functions in the heart, retina and central nervous system, and interacts with the hormonal system; its dietary importance as a true vitamin has been confirmed; in times of limited intake of methionine or cysteine and vitamin $B_6$ deficiency, taurine becomes essential; in addition, taurine is a strong physiological antioxidant[17].

Also *Opuntia ficus-indica* is a good source of betalains that are phytochemical compounds that play an important role in human health because of their pharmacological activities such as antioxidant, anti-cancer, anti-lipidemic and anti-microbial[23].

## EFFECTS OF *OPUNTIA* SPP. ON METABOLIC SYNDROME

Metabolic syndrome describes the increasing incidence of type 2 diabetes in association with obesity, hyperinsulinemia associated with disorders of metabolism of carbohydrates and lipids. Diabetes mellitus represents one of the biggest problems in public health; scientific research reported that consumption of *Opuntia ficus-indica* could reduce postprandial blood glucose, serum insulin, and plasma glucose-dependent insulinotropic peptide peaks, as well as increase antioxidant activity in healthy people and patients with type 2 diabetes; patients with type 2 diabetes who consumed the high-carbohydrate breakfast plus *Opuntia ficus-indica* showed a significantly lower area under the curve for glucose than those who consumed the high-carbohydrate breakfast only[8].

Studies demonstrated that the oral intake of nopal (*Opuntia* spp.) induces an hypoglycemic effect both in experimental animals and humans; the ingestion of *Opuntia* stems decreased the hyperglycemia induced by an oral glucose load in healthy subjects and was shown to be helpful in the short term treatment of diabetic patients including non insulin-dependent diabetes mellitus patients[9].

A scientific study showed that cactus fruit juice has a strong antioxidant property and can scavenge reactive oxygen species which resulted in markedly reduced hyperglycemia, hypercholesterolemia, lipid peroxidation, elevated levels of urea, improves the activities of antioxidant enzyme, blood reduced glutathione, serum aminotransferases, and increased the hemoglobin, protein and liver glycogen content in diabetic rats[11].

### Anti-hypercholesterolemia activity

Experimental evidence suggests that cactus pear reduces cholesterol levels in human blood and modifies LDL (Low-Density Lipoprotein) composition; the cholesterol, LDL and triglyceride plasma levels of rats were strongly reduced after 30 days of a daily administration (1 g/kg) of lyophilized cladodes of *Opuntia ficus-indica* L.[12]; also, several researchers have observed decreases in the LDL-cholesterol and triglycerides when individual dietary intakes were supplemented with prickly pads at 40–50 age-days.

*Opuntia ficus-indica* also is a source of betanin and indicaxanthin pigments with functional properties that can be used as food colorants. Betanin and indicaxanthin may be involved in improvement of the oxidative status of LDL in vivo; moreover, other studies showed that consumption of cactus pear fruit positively affected the body's redox balance and decreased the oxidative damage of lipids, and that the intake of red beet juice also delayed the LDL oxidation[22].

### Anti-cancer properties

Different wild to domesticated *Opuntia* species used as food resources and in folk medicine, exhibit antioxidant and cytoprotective biological properties that may potentially protect against the development of diseases in which oxidative stress and oxidized lipids are involved, such as colon cancer or atherosclerosis; scientific studies showed that colon and prostate cell lines were the most affected by the *Opuntia* juices, also a phytochemical agent isolated from the fruits of *Opuntia* was shown to exert an apoptotic effect on a human chronic myeloid leukemia cell line[13].

Scientific studies showed that the pigments betanin a betacyanin extracted from fruits of *Opuntia ficus-indica* have antiproliferative effects inducing apoptosis in human chronic myeloid leukemia cell line-K562[2]. Also other studies suggests that the cactus pear fruit extract inhibits the proliferation of cervical, ovarian and bladder cancer cell lines in vitro, and suppresses tumor growth in the nude mice ovarian cancer model in vivo[3].

Betalains from *Opuntia ficus indica* fruits were found to possess an anti-proliferative activity in cancer cells; prickly pear extracts containing betalains had a significant growth inhibitory effect on human ovarian cancer cells similar to a synthetic chemopreventive agent, 4-HPR, in nude mice; these extracts also inhibited the growth of 40-60% of immortalized cervical epithelium cells and cervical cancer cells; betalain (betanin) were also found to induce dose and time dependent apoptosis of the human chronic myeloid leukaemia cell line (K562) and decreased the membrane potential of mitochondria with simultaneous appearance of cytochrome c in the cytosolic fractions of cells; thus, a combination of drug treatment with betalains which stops cells at different cell cycle checkpoints (G1 and/or S phases) could be a promising strategy to inhibit tumour survival[23].

Cactus pear extracts, a natural product, has anti-cancer activity, although the active component(s) have not been clearly identified and as chemopreventive agents may need to be used for a long period of time to be effective and cactus has no toxic effects, cactus pear extracts can be easily used, for example, as dietary supplements in normal and high risk populations in chemoprevention[14].

The cactus pear extracts tested in this study could be such a candidate in cancer prevention for both normal and high-risk populations and prevention of recurrence of cancers[19].

## PREVENTION OF OSTEOSPOROSIS

Nopal could be an excellent complement of food in order to complement dietary nutritional intake and contribute to for the prevention of some osseous disorders such as osteoporosis[7]. For example, according to scientific results, nopal powder may prove to be an **alternative source of calcium** in populations for which the availability of dairy products is difficult and also for those who suffer from lactose intolerance[18].

It is well known that dietary calcium from vegetables is an excellent option to improve or maintain bone health; it was found that calcium intake, especially calcium from a vegetable source, such as nopal, is an **excellent nutraceutical to improve Bone Mass Density (BMD)**; the maximum deshydrated nopal dose used in this experiment improves BMD in women with low bone mass with hypercalciuria in the total hip and lumbar spine mainly in premenopausal women, maintaining the calciuria at the normal levels; the chemical proximate analysis of 600 g nopal used in this study showed that it contains a high amount of calcium and a low calcium oxalated content, which makes it an important source of dietary calcium, so deshydrated nopal can be considered as an excellent calcium source[16].

## PROSTATE ILLNESS

Cactus flower extracts have proved beneficial in benign prostatic hyperplasia by 5α-reductase and aromatase inhibition and exhibited anti-oxidative action[17].

As a conclusion, in view of the results of the scientific tests carried out, it can be said that *Opuntia* spp. is an excellent candidate for the development of multiple and effective phytopharmacological products with broad health benefits.

# References

[1] Ginestra, Giovanna et al. 2009. Anatomical, Chemical, and Biochemical Characterization of Cladodes from Prickly Pear [Opuntia ficus-indica (L.) Mill.]. J. Agric. Food Chem., 57, 10323–10330.

[2] Guedes Paiva, Patrícia Maria et al. 2016. Opuntia sp. Cactus: Biological Characteristics, Cultivation and Applications. AIR, 7(3): 1-14.

[3] Kaur, Manpreet et al. 2012 . Pharmacological actions of Opuntia ficus indica: A Review. Journal of Applied Pharmaceutical Science 02 (07); 15-18.

[4] Astello-García, Marizel G. et al. 2012. In Hispanic Foods: Chemistry and Bioactive Compounds. Chapter 8: Opuntia spp. as a Source of Bioactive Compounds. American Chemical Society.

[5] López-Cervantes, Jaime et al. 2011. Functional properties and proximate composition of cactus pear cladodes flours. Ciênc. Tecnol. Aliment., Campinas, 31(3): 654-659.

[6] El-Mostafa, Karym et al. 2014. Nopal Cactus (*Opuntia ficus-indica*) as a Source of Bioactive Compounds for Nutrition, Health and Disease. Molecules, 19, 14879-14901.

[7] Hernández-Urbiola, Margarita I. et al. 2011. Chemical Analysis of Nutritional Content of Prickly Pads (*Opuntia ficus indica*) at Varied Ages in an Organic Harvest. Int. J. Environ. Res. Public Health, 8, 1287-1295.

[8] Ondarza, Mauricio A. 2016. Cactus Mucilages: Nutritional, Health Benefits and Clinical Trials. Journal of Medical and Biological Science Research Vol. 2 (6), pp. 87-103.

[9] Frati, Albedo C. et al. 1990. Hypoglycemic Effect of *Opuntia ficus-indica* in Non Insulin-dependent Diabetes Mellitus Patients. Phytotherapy Research, VOL. 4, NO. 5.

[10] Santos Díaz, María del Socorro et al. 2017. Opuntia spp.: Characterization and Benefits in Chronic Diseases. Oxidative Medicine and Cellular Longevity.

[11] Hassan Abd El-Razek, Fatma and Hassan, Amal A. 2011. Nutritional Value and Hypoglycemic Effect of Prickly Cactus Pear (Opuntia Ficus-Indica) Fruit Juice in Alloxan-Induced Diabetic Rats. Australian Journal of Basic and Applied Sciences, 5(10): 356-377.

[12] Chauhan, S. P. et al. 2010. Biological Actions of Opuntia Species. Systematic Reviews in Pharmacy, Vol 1 | Issue 2.

[13] Keller, Julia et al. 2015. Antiatherogenic and antitumoral properties of Opuntia cladodes: inhibition of low density lipoprotein oxidation by vascular cells, and protection against the cytotoxicity of lipid oxidation product 4-hydroxynonenal in a colorectal cancer cellular model. J Physiol Biochem.

[14] Zou, Da-ming et al. 2005. Cactus pear: a natural product in cancer chemoprevention. Nutrition Journal, 4:25.

[15] Barba, Francisco J. et al. 2017. Impact of conventional and non-conventional processing on prickly pear (*Opuntia* spp.) and their derived products: From preservation of beverages to valorization of by-products. Trends in Food Science & Technology 67, 260-270.

[16] Aguilera-Barreiro, Marıa de los Angeles et al. 2013. Intake of dehydrated nopal (*Opuntia ficus-indica*) improves bone mineral density and calciuria in adult Mexican women. Food and Nutr. Research, 57: 19106.

[17] Stintzing, Florian C. et al. 2001. Phytochemical and nutritional significance of cactus pear. Eur Food Res Technol, 212 : 396–407.

[18] Contreras-Padilla, Margarita et al. 2011. Evaluation of oxalates and calcium in nopal pads (Opuntia ficus-indica var. redonda) at different maturity stages. Journal of Food Composition and Analysis 24 38–43.

[19] Toit, Alba du et al. 2018. Cultivar and Harvest Month Influence the Nutrient Content of Opuntia spp. Cactus Pear Cladode Mucilage Extracts. Molecules, 23, 916.

[20] Pavithra K. et al. 2015. Pectin and Isolated Betalains from *Opuntia dillenii* (Ker-Gawl) Haw. Fruit Exerts Antiproliferative Activity by DNA Damage Induced Apoptosis. IJPPR, Volume 7, Issue 6.

[21] De Waal, H.O. et al. 2015. Development of a cactus pear agro-industry for the sub-Sahara Africa Region. Proceedings of International Workshop, University of the Free State, Bloemfontein, South Africa, 27-28, 96 pp.

[22] Sáenz, Carmen et al. 2012. Red betalains from Opuntia spp.: Natural colorants with potential applications in foods. Israel Journal of Plant Sciences, Vol. 60 pp. 291–299.

[23] Gengatharan, Ashwini et al. 2015. Betalains: Natural plant pigments with potential application in functional foods. LWT - Food Science and Technology 64, 645e649.

[24] Hernández-Urbiola, Margarita I. 2011. Chemical Analysis of Nutritional Content of Prickly Pads (*Opuntia ficus indica*) at Varied Ages in an Organic Harvest. Int. J. Environ. Res. Public Health, 8.

# CHAPTER 5

# ECOLOGICAL, BIOTECHNOLOGY AND INDUSTRIAL APPLICATIONS

# ECOLOGICAL, BIOTECHNOLOGY AND INDUSTRIAL APPLICATIONS

"The Tree of Life and The Green Gold" can easily grow without care or with very little care in arid and semi-arid zones and sustain life in adverse life zones by producing products that can contribute to develop and maintain communities in inhospitable regions. It is a source of food all year round both for humans and animals and it perform a major environmental role protecting slopes against erosion and flooding, as well as rehabilitating marginal lands by improving the levels of humidity and soil retention capabilities. Also it allow important socioeconimic industry developments.

## INTRODUCTION

*Opuntia* is a perennial plant that should be called "The Tree of Life and The Green Gold". The physiological and morphological characteristics of nopal plants allow them to survive in dry environments and in soils with low fertility where other crops cannot thrive[4].

Cacti are the most conspicuous and characteristic plants of warm arid regions; cacti are remarkable for their diversity of growth forms and their ability to not only grow, but to thrive under environments recognized as stressful for most plant species; cacti can be used to prevent soil erosion and act as an effective living fence[15].

*Opuntia* scrublands perform a major environmental role protecting slopes against erosion and flooding, as well as rehabilitating marginal lands by improving the levels of humidity and soil retention capabilities; *Opuntia* scrubs are also used for animal grazing all year round and can become an emergency feedstock in case of drought[1].

*Opuntias* can endure heavy and continuous pruning; the plant can be maintained in the juvenile stage indefinitely by continuous pruning, which is the basic crop management tool for vegetable *Opuntia* production; application of high manure rates to soil and pruning are responsible for the high yields observed in vegetable production in Milpa Alta, Mexico, which can reach 400 ton/ha/year[2].

The agricultural activity of prickly pear orchards, in addition to avoiding damages to the productive ecosystem also generates socio-economic benefits; also the prickly pear is a plant that is grown in orchards, and it does not compete with other local crops: maize, barley, wheat and beans, but instead compliments the basic agricultural activities which is why it is largely grown on hillsides or on flat land unsuitable for conventional crops[6].

Cactus (*Opuntia ficus indica*) is a long-domesticated cactus crop that has emerged as one of the most suitable crop species for arid and semi-arid climate for not only for its better Water Use Efficiency (WUE) and Rain Use Efficiency (RUE) but also a source to meet the requirements of food, forage and host of other benefits along with ecological advantages[11].

Also, in Central Africa, the sap from the pads served as a mosquito repellent; in 1911, Burbank noted in Scientific American, that when spread on water, it smothers mosquito larvae, and the effect lasts up to a year[18]. This discovery is another apparent and beneficial effect that mucilage could have to control mosquito proliferation without causing damage to ecosystems, so this property deserves to be investigated and studied in depth.

## "THE TREE OF LIFE AND THE GREEN GOLD" AS FORAGE AND AN USEFUL PLANT FOR FEEDING ANIMALS

Within the genus *Opuntia*, *Opuntia ficus-indica* is the most important for its production of edible fruits and cladodes, which are used as a vegetable and as animal forage[15].

*Opuntia* spp. is particularly attractive as a feed because of its efficiency in converting water to dry matter, and thus to digestible energy. *Opuntia* spp., have been extremely useful livestock forage in times of drought, primarily by providing digestible energy, water and vitamins. Although mainly used for cattle, *Opuntia* has also been used as forage for pigs; however, it must be combined with other foods to complete the daily diet because *Opuntias* spp. are poor in proteins, although rich in carbohydrates and calcium[3]. Growing *Opuntia* for forage production needs careful timing of pruning practices[2].

Some livestock species such as the goat already graze the cactus in certain parts of India, particularly in dry and arid areas; there the cactus is used to feed livestock and small ruminants such as goats by mixing it with other foods that become part of the Mixed Total Ration (MTR); many farmers are encouraged to cut the cactus into smaller pieces and supplement it with hay or straw; the cladodes that form the segmented stems of the cactus store water and are rich in easily fermentable carbohydrates and aid digestion in the rumen; goats usually consume an average of 3 kg of cactus per day and can maintain their body weight without any adverse effect on their health; the cactus can be used as fodder to replace 25% of the dry matter corresponding to regular grazing[11].

*Opuntia* spp., a succulent and drought-tolerant species, can produce more than 20 tonnes of dry matter per hectare per year and provide 180 tonnes per hectare per year of water stored in its cladodes which represents a cost-effective option for livestock water supply; cactus cladodes and fruit wastes are also a cost-effective feed for ruminants. The benefits of using this cactus as feed are well documented[12], although cladodes such as fodder can solve the problem of water supply for livestock, attention must be paid to its low dry matter content in relation to diet composition; to compensate for the low dry matter content, the ruminant consumes large amounts of cladodes, which can cause diarrhoea; therefore, it is recommended that other appropriate supplementary feeds be added to the animals' diet, particularly those that are rich in nitrogen[12].

Cladodes were used as supplementary feed to feed different types of animals such as small goats and lambs; the result was positive for dry matter intake, total live weight gain and feed conversion and profitability; lambs fed a high proportion of cladodes were more efficient in converting feed into total live weight gain, so cactus can offer an alternative to traditional feed for lambs without affecting the productivity performance of the animals and at the same time reducing production costs[15].

Some of the developments aimed at improving the production of nopalitos can be adapted to similar forage production systems because *Opuntia* spp. is a plant that tolerates competition and intensive pruning[2].

## CACTUS IN CROPPING SYSTEMS

Expansion of cereal cropping into rangelands and reduction in fallow practice are among the main reasons for declining soil fertility and wind erosion; one way of combating degradation resulting from cereal monocropping is the introduction of adapted forage legumes, fodder shrubs/trees and cactus in cropping systems; alley cropping is an agroforestry practice where perennial crops are simultaneously grown with an arable crop such as barley, oat etc.; alley cropping overcomes some of these constraints because it improves soil fertility, increases crop yield, reduces weeds, and improves animal performance; properly managed alley cropping allows diversification and farmers can benefit from several markets; it also promotes sustainability in both crop and livestock production by increasing land productivity and reducing weather risks compared with cultivation of annual crops only[12].

Another way to improve and to increase the *Opuntia* spp. production, it is cultivating in the prickly cactus field, *Trifolium pratense* or red clover, a leguminous that is a beneficial plant that can fix nitrogen increasing the fertility of the soil.

As it is well known, soil nitrogen (N) is a mayor limiting factor for plant production in most ecosystems of the world; among $N_2$ fixing plants, forage legumes have the advantage that a larger proportion of the nitrogen accumulated generally is available for return to the soil through leaf fall, grazing animals an root turnover than in grain legumes where much nitrogen is harvested in the seed; red clover, a widely distributed perennial legume found in meadows and pastures throughout Europe and elsewhere, is also an important forage and green manure crop world-wide[13]. Red clover is a forage legume cultivated in the temperate world, noted for its high-protein feed and high rate of biological nitrogen fixation and the use of red clover in soil fertility-building could be an useful contribution to the soil because it can make to sustainable intensification and the reduction of agriculture's environmental impact[14].

## AGRO-INDUSTRIAL UTILIZATION

The industrial opportunities for *Opuntia* extend beyond food products, making the species valuable for both cropping and utilization[16]. "The Tree of Life and The Green Gold" is natural resource that allow economic development with important economic improvement for farmers. It can be a primer matter source both for food and not food industry.

Dye production from cochineal cultivation and phytochemical betalains, flour and biofloculants production are some of the agro-industrial products that can be obtained from "The Tree of Life and The Green Gold".

Also, food industry can take advantage of the physicochemical properties of fiber to improve the viscosity, texture, sensory characteristics and shelf-life of their products[18].

### COCHINEAL AND CARMINIC ACID COLORANT PRODUCTION

One of the more attractive indirect products made from *Opuntias* is carmine from the cochineal insect (*Dactylopius coccus*); this is the best and safest natural red colouring agent for foods;

carmine, carminic acid or solutions; the cochineal is an insect of the Hemiptera order and parasitic to the *Opuntias* on which it feeds; the plants must be managed differently from those used for the production of fruit or cladodes; in addition, the insect has to be cultivated, harvested and processed to produce the colorant[16].

Cochineal is using for production of dye; the dye extracted from cochineal *(Dactylopius coccus)* is rich in carminic acid and used for coloring of food and cosmetics, textiles and drugs[5].

Cochineal is a sessile parasitic insect living on *Opuntia* plants; first instar nymphs are the only mobile life stage; these nymphs crawl and disperse along the host plant searching for a place to settle or are transported by the wind to other potential hosts, thus colonizing distant areas; once settled, nymphs become sessile and develop to adult insects; due to their lack of mobility, cochineals are easily collected by peasants using soft brushes or spoons to withdraw them from their host plants; in most cases, only the less accessible insects or plants are left untouched[1].

Cochineal life cycle varies according to their sex; the female life cycle has two nymph and one adult stage, whereas the male passes through four nymph stages; the first two instars being comparable to that of the female but the third and fourth are characterized by a pre-pupa and pupa stage; additionally, female cochineal metamorphosis is incomplete whereas male is complete metamorphosis; the life cycle of cochineal varies based on climatic variables such as temperature and it ranges from 90 to 128; cochineal breeding is affected by temperature rain; female cochineal mostly lay up to 420 eggs[5]. The mean number of eggs produced in an ovisac is 130 per female; these remain in the abdominal side of the female; the adult females remain attached to cladodes while males are winged and have a shorter life of 7 days; *Dactylopius opuntiae* is used as a natural enemy in countries where the cactus is considered a weed; only females and nymphs feed on plants and cause damage by sucking sap from the cladode; carminic acid produced by *Dactylopius opuntiae* is involved in protection of the insect from natural enemies[7].

One fascinating use of *Opuntia ficus-indica* was as a host plant for the growth of the cochineal insect (*Dactylopius* spp.) which is the source of a highly appreciated red dye such as carminic acid; cochineal dye was highly valued and became the third greatest source of income for the Spanish crown in the New Spain, only after gold and silver; the successful production of cochineal dye increased its demand worldwide both in the European and Asian markets the expansion of *Opuntia* species and cochineal production continued during the XVII and XVIII centuries[4].

Another example of cochineal business was the introduction of the carmine cochineal to Ethiopia was mainly for export purposes because there was no knowledge about its use in Ethiopia; new cactus plantations started and in a time span of 2-3 years the company developed 135 ha of the pledged 300 ha with new plantations of cactus pear that were infested with cochineals; within a short period of time the company started harvesting carmine cochineals and export of dried cochineal insects; the cochineal business created job opportunity for 32 full time workers and more than 136 casual labourers; as cochineal is an introduced insect, there are no natural enemies that can infest cochineal in Ethiopia; so this was a very good opportunity for the expansion of the cochineal business in the region[8].

## Betalains pigments from fruits as colorants

Betalains are pigments found in red and purple cactus pear fruit that are widely used in the food industry[16].

*Opuntia* have several commercial betalain, a water-soluble nitrogen-containing pigment found in high concentrations in cactus pear plants can be used as natural food colouring agent[9]. Betalain pigments are bioactive compounds with diverse biotechnological applications, they are conformed by two main groups: betacyanins (red) and betaxanthins (yellow), which could have a great industrial and food applications because they could replace artificial dyes that cause allergic disorders and are carcinogenic[4].

### Seed oil

The *Opuntia* seed oil is obtained by cool pressed seeds and some of its main applications are being developed by the cosmetic industry; the seed oil is destined for cosmetic product production, and it sells at a very high price as organic oil for anti-age and anti-wrinkle purposes[10].

### A source of natural sweeteners

The use of natural sweeteners other than sucrose is an interesting area for the food industry; the use of high fructose corn syrup (HFCS) as a sweetener has increased in recent years; the food industry uses these sweeteners for beverages, cookies, ice-cream and other products; the results of the analysis done indicated the presence of glucose and fructose; the amount of glucose (57.2%) was confirmed by the analysis of optical rotation (57.1% glucose and 42.9% fructose); cactus fruit can be used as a raw material in the creation of natural sweeteners[17].

## BIOTECHNOLOGY

Biotechnology, through research and application, upgrades the biological raw materials of *Opuntia* creating new value streams, whereas the cultivation of nopal bioresource promotes numerous ecological services including promotion of the biodiversity, prevention of soil degradation and water supply in arid and semi-arid areas (bio-ecology); whatever the bioproduct or the mixture of bioproducts derived from *Opuntia*, today's and tomorrow's nopal-based bioeconomy companies will adopt clean production technologies[8]. Tissue culture can be an excellent tool for cacti micropropagation; many cacti produce an excess of auxin under in vitro culture conditions, and this stimulates callus production; advantages of the propagation systems outlined above include the time required for production of cacti shoots by microprogagation is quite short[15].

## BIOGAS FOR A FRIENDLY ENVIRONMENT

In Mexico, for instance, since 2016 a tortilla making company (NopaliMex) uses cladode-based biogas to meet all its heat, fuel and electricity needs; in detail, the company uses two closed biodigester tanks where cladodes from its 70 ha cactus plantation are biodigested anaerobically to produce biogas and fertilizers (nitrogen-rich water and humus); the cladodes are cut before inserting them into the biodigestor in large tanks with water at 38°C where bacteria decompose the mixture in high quality biogas (75% methane, 24% carbon dioxide and 1% minor gases and no trace of hydrogen sulfide); the heating capacity of the biogas is 7,000 kcal/m$^3$ (8,800 kcal/m$^3$ for 100% methane); further treatment of the biogas thereby obtained with sulphuric acid extracts carbon dioxide, giving 96% biomethane which is used to fuel retrofitted internal combustion engine vehicles[4].

## THE BENEFITS OF AGRICULTURAL SUSTAINABLE ECOSYSTEMS

Agricultural ecosystems sustainability depends on the ability of farmers to maintain soil productivity managing the biodiversity to produce food and at the same time looking for an efficient agricultural management that benefit the ecosystem.

In agricultural systems, biodiversity performs ecosystem services beyond production of food, fiber, fuel, and income; examples include recycling of nutrients, control of local microclimate, regulation of local hydrological processes, regulation of the abundance of undesirable organisms, and detoxification of noxious chemicals[20].

Agricultural biodiversity includes all components of biological diversity relevant to the production of goods in agricultural systems: the variety and variability of plants, animals, and microorganisms at genetic, species, and ecosystem levels that are necessary to sustain key functions, structures, and processes in the agroecosystem and also it includes crops, trees, and other associated plants, fish and livestock, and interacting species of pollinators, symbionts, pests, parasites, predators, and competitors[19].

Multiple cropping systems use management practices where the total crop production from a single piece of land is achieved by growing single crops in close sequence, growing several crops simultaneously, or combining single and mixed crops in some sequence; in areas of the world where multiple cropping is a common aspect of agroecosystem management, productivity generally is more stable and constant in the long term[21].

It is important to manage the appropriate introduction of combination of species for getting agricultural production with profit thinking in systems that interact benefiting the agricultural ecosystem avoiding as much as possible the input of agrochemicals such as fertilizer, pesticides and herbicides. Inappropriate or excessive use of some inputs often reduces biodiversity in agricultural ecosystems (thus compromising future productivity) and in other ecosystems[19].

Intercropping, agroforestry, shifting cultivation and other traditional farming methods mimic natural ecological processes, and their sustainability lies in the ecological models they follow; this use of natural analogies suggests principles for the design of agricultural systems that make effective use of sunlight, soil nutrients, rainfall, and biological resources[20].

"The Tree of Life and The Green Gold" is a plant that can be managed with other crops in an efficient, productive and sustainable agricultural ecosystem.

# References

[1] Rodríguez, Luis C. et al. 2006. Local identification and valuation of ecosystem goods and services from Opuntia scrublands of Ayacucho, Peru. Volume 57, Issue 1, Pages 30-44.

[2] Mondragón-Jacobo, Candelario et al. 2001. Cultivation of opuntia for fodder production: from re-vegetation to hydroponics. FAO.

[3] Reynolds, Stephen G. and Arias, Enrique. 2001. General background on *Opuntia*. Introduction. Opuntia as forage. FAO.

[4] Ciriminna, Rosaria et al. The unfolding bioeconomy of nopal (*Opuntia* spp.).

[5] Abrha, Haftu et al. 2018. Modeling the impacts of climate change and cochineal (*Dactylopius coccus* Costa) invasion on the future distribution of cactus pear (*Opuntia ficus-indica* (L.) Mill.) in Northern Ethiopia. JPACD, 20:128-150.

[6] Losada, Hermenegildo R. et al. 2017. Economic Indicators, Capacity of the Ecosystem of Prickly Pear Cactus (*Opuntia Megacantha*) and Environmental Services in Teotihuacan, México to Supply Urban Consumption. Journal of Agriculture and Environmental Sciences, Vol. 6, No. 1, pp. 85-91.

[7] Bouharroud, R. et al. 2018. Biological control of the prickly pear cochineal Dactylopius opuntiae Cockerell (Hemiptera: Dactylopiidae). Bulletin OEPP/EPPO Bulletin, 0 (0), 1–7.

[8] Belay, Tesfay. 2015. Carmine cochineal: fortune wasted in northern Ethiopia. JPACD, 17:61-80.

[9] Pavithra K. et al. 2015. Pectin and Isolated Betalains from *Opuntia dillenii* (Ker-Gawl) Haw. Fruit Exerts Antiproliferative Activity by DNA Damage Induced Apoptosis. IJPPR, Volume 7, Issue 6.

[10] De Waal, H.O. et al. 2015. Development of a cactus pear agro-industry for the sub-Sahara Africa Region. Proceedings of International-

al Workshop, University of the Free State, Bloemfontein, South Africa, 27-28, 96 pp.

[11] Kauthale, Vitthal et al. 2017. Cactus an Emerging Fodder Crop of Arid and semi Arid India. BAIF Development Research Foundation.

[12] FAO. 2017. Crop ecology, cultivation and uses of cactus pear. Food and Agriculture Organization of the United Nations and the International Center for Agricultural Research in the Dry Areas.

[13] Warembourg, F. R. et al. 1997. Economy of symbiotically fixed nitrogen in red clover (*Trifolium pratense* L.). Annals of Botany 80:515-523.

[14] McKenna, Patrick et al. 2018. The use of red clover (*Trifolium pratense*) in soil fertility-building: A Review. Field Crops Research 221: 8–49.

[15] Shedbalkar, Utkarsha U. et al. 2010. *Opuntia* and Other Cacti: Applications and Biotechnological Insights. Tropical Plant Biol. 3:136–150.

[16] Sáenz, Carmen et al. 2013. Agro-industrial utilization of cactus pear. Food and Agriculture Organization of the United Nations, FAO.

[17] Sáenz, Carmen et al. 1998. Cactus pear fruit: A new source for a natural sweetener. Plant Foods for Human Nutrition 52: 141–149.

[18] University of California. 1989. Prickly Pear Cactus Production. Small Farm Center, University of California.

[19] Jarvis, D. I. et al. 2007. Managing Biodiversity in Agricultural Ecosystems. Bioversity International.

[20] Altieri, Miguel A. 1999. The ecological role of biodiversity in agroecosystems. Agriculture, Ecosystems and Environment 74: 19–31.

[21] Gliessman, Stephen R. Capter V: Multiple Cropping Systems: A Basis for Developing an Alternative Agriculture. University of California.

Made in the USA
Monee, IL
05 November 2026

67383110R00080